營養師給孩子的36道有營料理

——讓孩子從此愛上吃！

註冊營養師
陸蕙華（Denise Luk）
盧庭威（Kurtus Lo） 著

萬里機構

推薦序 1

Denise 是我大學碩士班的同學，其後更成為我工作上的夥伴、好友。一年前，當我得知她在百忙的工作之中生起出版這本書的念頭時，心裏既欣賞她的熱誠，又替她擔心能否完成這份艱巨的「任務」。

因為自身工作的關係，時常需要接觸關於兒童的營養飲食，外國的網站提供了很多詳細的資料甚至食譜，但往往不太符合本地人的口味。本地製作的食譜通常都頗成人化，它們的營養價值或者不太符合小朋友的需要。

我身邊的同事以至家長們時常希望有一本有趣的書，能夠從小朋友的營養價值入手，甚至可以透過帶動大小朋友的參與而令製作出來的食物更加「美味」！

這本《營養師給孩子的 36 道有營料理——讓孩子從此愛上吃！》最終出版了！它是專門為喜歡親子烹飪的爸爸、媽媽而寫，營養師們精心設計的食譜既注重兒童的營養價值，有部分更加針對小朋友常見的健康問題，例如偏食、便秘等，間場的「營養師聊天室」更可以透過問答環節增加互動性。本書內容有趣、淺顯明白，又可以鼓勵不同年齡層的大小朋友「落手落腳」幫忙，增進親子關係；享受各式各樣美食的同時，也能夠學習關於飲食和烹飪的知識。

感謝 Denise 和 Kurtus 為喜歡親子烹飪的爸爸、媽媽編寫了一本健康、有趣的好讀物，亦感謝每一位讀此書的大小朋友，希望大小讀者們喜歡這本書之餘，也能令你們食得更有「營」。

傅振祥醫生

兒童內分泌科專科醫生

中文大學兒科學系榮譽副教授

香港大學兒科學系榮譽助理教授

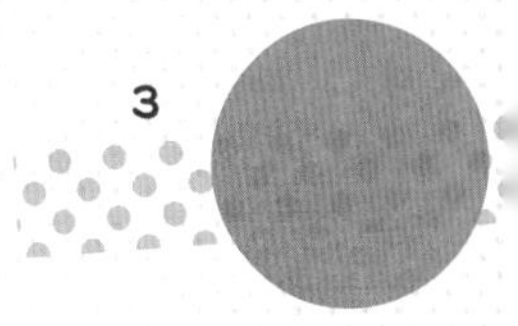

推薦序 2

親子食譜又關我事？

冇錯！我嘅口味同喜好其實好小朋友，朱古力、薯片、爆谷無一不歡，家長們眼中嘅「垃圾食物」全部我都好喜愛，所以我絕對有資格用小朋友嘅角度去分享呢本書。

寫呢個序時，我都有諗點解我咁喜歡「垃圾食物」呢？回想我小時候，媽媽對我極為嚴格，成日禁止我吃糖果同薯片，話唔健康嗎！所以，餐枱上出現最多嘅顏色就係綠色同白色嘅食物，唔係蒸煮就係白烚嘅菜式。到大個咗，香口食物就不知不覺咁成為咗我嘅任性減壓食物啦。

其實，我都想食健康啲㗎，但身為地獄廚神嘅我，只好搵下有冇啲又易整又好味嘅食譜呢？估唔到，搜羅咁多食譜之後，竟然發現呢本親子食譜最啱我！佢整法簡單、營養豐富，而且顏色又靚，仲可以順便「打卡」，呵呵！

至於各位家長，呢本書仲可以一書幾用，同小朋友討論整邊樣菜式時，可以了解佢地嘅口味同喜好；去買食材時，可以教導佢地認識唔同食材嘅名同營養價值；準備材料時又可以學埋計算。而最重要梗係一齊整可以增進感情啦！

另外，私心介紹呢本書，完全喺因為認識咗好幾年嘅營養師 Kurtus，每年我哋都會為香港中文大學舉辦嘅糖尿病公眾講座做嘉賓同烹飪示範。糖尿病明明喺飲食方面有較多嘅限制，但 Kurtus 都可以設計出好多病人都話估唔到可以食又符合病情要求嘅食譜。Kurtus 好明白美食可以治癒人心，只要計得好，煮得精明，就可以食得有營。

今次 Kurtus 同另一位營養師 Denise 為大家製作咗 36 道親子營養手作料理，下本就要 72 變啦！期待將來有更多創新嘅健康食譜畀我地大人細路，等大家由細到大都可以「食得營、變得型」！

梁嘉琪
藝人

序

家長們總希望把最寶貴的財富——「健康」送到子女身上，更希望在孩子成長中把那份「親子回憶」刻進家人的心中。筆者認為一本成功的親子食譜，除了提供美味健康的食物，更應為每位讀者與家人共同製造溫馨的回憶，而這份「親子回憶」也正是不少家庭將來成為獨自擁有的「家傳食譜」了。

現實中，子女為着學業、他人的期望而努力；家長則為供應最好的東西給子女而默默付出，忙碌的生活中彷彿少了一份互動的樂趣。筆者希望透過親子烹飪活動，令家長和子女的關係拉近。

書中介紹多款親子料理，除「營養健康」、賣相「吸睛」、「打卡一流」外，當中更附有親子步驟，讓家長和子女一起製造美味回憶。不少家長擔心子女烹調食物時會受傷，但製作過程正是讓子女學習的好機會，除了從中認識健康食材，更可學習基本烹飪技巧和訓練手部活動，當中與子女的溝通和互動更是重要的過程。製作過程中少不免會出錯，製成品或許未如預設圖一樣，但這一切正是屬於你們的親子回憶，亦是教導子女如何從錯誤中汲取經驗和改善的機會。在此，筆者希望家長閱讀此書時不妨放鬆心情，以享受的心境與子女共同製作料理。

食物不僅能提供營養，更為人與人之間建立聯繫，是家庭、文化和傳統的核心。希望讀者可以多了解和尊重食物，與食物慢慢培養出良好關係，終身受用。

藉此再次多謝出版社、兩位插畫師 Janice Tam 與 Nicole Luk 友情客串、營養科學生 Keung Yan Wah 及 Adele 的幫忙。還有來自三個家庭的家長和未來小廚神參與拍攝——Jess 媽媽與女兒 Janis、Betty 媽媽與兒子 Je Hyun、Tina 媽媽與子女 Raina、NokNok，以及所有支持此作品的朋友們。

陸蕙華（Denise Luk）
盧庭威（Kurtus Lo）

作者簡介

陸蕙華（Denise Luk）

註冊營養師及糖尿教育者專業資格（加拿大）
香港營養師協會培訓及發展主任、媒體發言人（2019-2021）及正式會員

加拿大麥基爾大學營養學學士（飲食治療）及中文大學內分泌及糖尿治理理學碩士。修畢澳洲 FODMAP 飲食培訓課程、美國波士頓大學兒科營養研究生課程。

曾於加拿大大型醫院及老人院提供營養師服務；回港後，於國際及本地非牟利機構擔任社區營養師，致力推廣健康生活和疾病預防的重要性，為不同範疇的學校、教育學院、志願機構和大型企業策劃和主持營養工作坊、健康烹飪班和課程，為不同年齡層、低收入家庭、少數族裔人士提供個別營養諮詢和服務。擁有多年為備孕／懷孕及母乳餵哺、體重管理、糖尿病、高血壓、高血脂、痛風、脂肪肝、腸胃問題、腎病和癌症等相關人士等提供專業飲食治療的經驗，更為多間教育學院擔任兼職講師。

曾與多個機構協力製作健康資訊刊物，經常獲邀接受電台和電視台（如無綫電視、有線電視、NOWTV、香港電台等）和多個社交平台的傳媒專訪，定期為多間本地報章、雜誌和會刊專欄撰寫文章，從多方面推廣健康飲食。

盧庭威（Kurtus Lo）

註冊營養師（澳洲）、運動營養師（澳洲）
香港營養師協會財政、媒體發言人（2019-2021）及正式會員

畢業於澳洲紐卡素大學膳食及營養學學士課程（一級榮譽）及澳洲運動營養學會課程，修畢香港大學公共衞生碩士課程。

現職主要推廣健康飲食計劃，於政策層面改善大眾健康，並於教育學院擔任兼職講師，期望教育及培養大眾對公共衞生營養的興趣。經常於電台及電視不同傳播媒體及研討會，分享健康飲食心得及烹飪示範，如商業電台、蘋果日報、明報、經濟日報、成報、晴報、AM730 之訪問，亦為無綫電視、有線電視及香港電台等資訊性節目作專業嘉賓分享。多年來為市民舉辦講座及個別諮詢服務，曾為低收入家庭、特殊兒童、少數族裔及不同年齡層人士推廣健康飲食。

他熱衷寫作、烹飪及創作料理，協助編著《水果營養全書》及《蔬菜營養全書》，定期於多份報章撰寫營養飲食主題文章。曾於米芝蓮餐廳當學廚，亦分別為本港主題樂園及香港營養師協會設計健康飲品及食譜，將營養理論實踐為生活一部分。

目錄

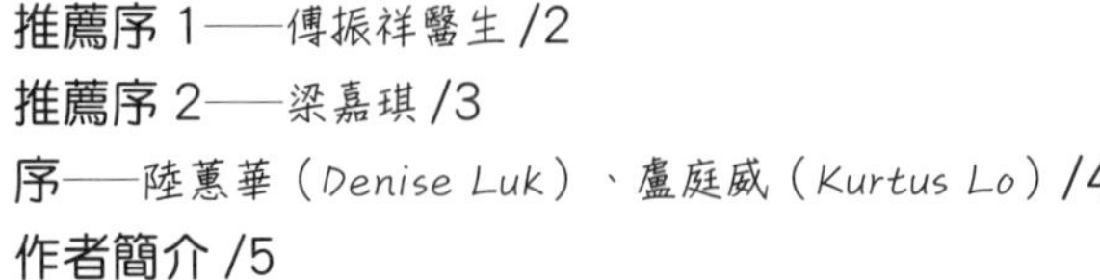

動動手，變出得意造型！

輕怡甜點

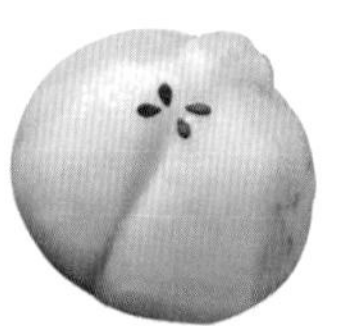

喜慶節日食品

Chapter 1

吃得好，學童夠精叻！

你是否有相同的疑問

子女是否過瘦？

孩子好像沒有長高？

同等年紀，她的身形比哥哥那時候還要細小……

鄰家同年紀的小朋友比自己的子女高出半個頭！

1.1 如何評估子女是否健康？

要了解子女的健康狀況，當然要準確地量度身高和體重。

量度身高的方法

❶ 準備量度工具，確保工具運作正常。
❷ 脫掉鞋子、除去帽及頭上飾物。
❸ 直立時，背部靠着牆身或垂直尺，雙腳伸直緊合。
❹ 頭部、背部、臀部、小腿和腳踭靠着牆身，放鬆身體。
❺ 眼望前方，視線保持水平位置。
❻ 以水平方向量度高度，以厘米作記錄單位。
❼ 重複量度，以兩次的平均數為準。

量度體重的方法

❶ 準備量度工具，確保工具運作正常。
❷ 體重磅放置於平坦地面，調整至「零」開始。
❸ 脫掉鞋子和厚身衣物，提醒孩子把玩具或褲袋的物件暫時取出。
❹ 保持身體平穩，雙手放於兩側，放鬆身體站立在體重磅上。
❺ 待量度數字穩定後，以公斤作記錄單位。

家長較常擔心子女是否健康成長，量度身高和體重後可利用以下圖表，查看子女的身高和體重是否屬於健康範圍吧！

注意事項

1. 密切留意男、女的不同表格。
2. 低於參考指數下限為過輕（體重低於身高別體重指數中位數的 80%）。
3. 超過參考指數上限為超重（體重高於身高別體重指數中位數的 120%）。

身高（厘米）	男：健康體重範圍（公斤）	女：健康體重範圍（公斤）
75	7.5-11.3	7.3-11.0
77	7.9-11.8	7.7-11.6
79	8.2-12.3	8.1-12.2
81	8.6-12.9	8.5-12.7
83	8.9-13.4	8.9-13.3
85	9.3-13.9	9.2-13.9
87	9.6-14.4	9.6-14.4
89	10.0-15.0	10.0-15.0
91	10.3-15.5	10.3-15.5
93	10.7-16.1	10.7-16.0
95	11.1-16.6	11.0-16.5
97	11.5-17.2	11.4-17.1
99	11.9-17.8	11.8-17.6
101	12.3-18.4	12.1-18.2
103	12.7-19.1	12.5-18.8
105	13.2-19.7	12.9-19.4
107	13.6-20.4	13.3-20.0
109	14.1-21.2	13.8-20.7
111	14.6-21.9	14.2-21.3

身高（厘米）	男：健康體重範圍（公斤）	女：健康體重範圍（公斤）
113	15.2-22.8	14.7-22.1
115	15.7-23.6	15.3-22.9
117	16.3-24.5	15.8-23.7
119	17.0-25.5	16.4-24.7
121	17.6-26.5	17.1-25.7
123	18.4-27.5	17.8-26.7
125	19.1-28.6	18.6-27.9
127	19.9-29.8	19.4-29.1
129	20.7-31.0	20.2-30.4
131	21.6-32.3	21.2-31.7

參考資料：

Leung SSF, Lau JTF, Tse LY, Oppenheimer SJ (1996). " Weight-for-age and weight-for-height reference for Hong Kong children from birth to 18 years." *J Paediatr. Child Health 32*(2),103-109.

給家長的話

營養師聊天室

每位學童成長的速度不一，遺傳、出生時的身長和體重、個別成長速度、營養攝取、運動程度等因素均影響成長，因此家長毋須過分擔心或與他人作比較。恆常地為子女量度身高和體重，作定期檢測其成長速度是否正常、體重是否符合健康範圍便可。如成長速度延緩引致體重過輕或過快導致過重時，應諮詢醫生或註冊營養師的意見。

1.2 想子女健康成長，食物重「量」重「質」

如何重「量」？

家長應培養子女實踐多元化飲食模式，並從不同種類食物攝取足夠的營養素。嘗試每天進食以下六種食物組別，並飲用足夠的流質飲品。由於不同年齡層所需的熱量和營養素要求不一，應按照子女的年齡而訂立其膳食分量。

穀物類

食物例子：全穀物及其製品、米飯、非油炸麵條（意粉、通粉、烏冬、米粉、米線、蛋麵、上海麵）、粥、麵包、麥皮、低糖早餐穀物脆片。

主要營養素：碳水化合物、維他命 B_1 和 B_6、植物性蛋白質。

營養素	作用
碳水化合物	於體內轉化為血糖，供能量給大腦和身體使用。
維他命 B_1	協助碳水化合物和蛋白質食物於體內分解並釋放能量。 保持神經系統、心臟、消化系統正常運作。
維他命 B_6	協助身體使用和儲存從碳水化合物和蛋白質食物中攝取的能量。 幫助製造血紅蛋白，有助紅血球運送氧氣和養分供身體使用。
植物性蛋白質	蛋白質的一種，有助身體建立肌肉、促進生長發育和修補身體組織。

蔬菜及水果類

食物例子（蔬菜類）：綠葉蔬菜（菜心、白菜、芥蘭）、瓜類（冬瓜、節瓜、翠玉瓜）、菇菌類（大啡菇、蘑菇、雲耳、金菇）、其他（秋葵、燈籠椒、番茄、椰菜花）。

食物例子（水果類）：瓜果類（西瓜、蜜瓜、哈密瓜）、柑橘類（橙、西柚、柚子）、莓類（士多啤梨、藍莓、紅莓）、梨果類（蘋果、啤梨、枇杷）、熱帶水果（香蕉、菠蘿、熱情果、火龍果）、無添加糖分乾果。

主要營養素：碳水化合物（根莖類蔬菜和水果）、膳食纖維、維他命 C、葉酸、礦物質（如鉀質）、抗氧化物、水分。

營養素	作用
碳水化合物（根莖類蔬菜和水果）	於體內轉化為血糖，供能量給大腦和身體使用。
膳食纖維	有助增加飽肚感，控制食量，有助體重管理。 非水溶性纖維：有助腸臟蠕動，預防便秘。 水溶性纖維：有助穩定血糖和膽固醇水平。
維他命 C	保護細胞完整性、幫助傷口癒合、協助鐵質吸收。
葉酸	協助身體製造紅血球，預防貧血。 減少胎兒患上神經管缺陷（如：先天腦部或脊椎發育異常）。
鉀質	平衡體內水分和電解質。 協助心臟肌肉正常運作。 攝取足夠有助控制血壓，減低中風的風險。
抗氧化物	擊退體內的自由基，減低患上癌症和其他慢性疾病的風險。

嘗試彩虹飲食法 Eat a Rainbow ！

建議家長可為兒童提供並鼓勵攝取不同顏色的蔬果，均衡攝取各種植物元素並豐富食物色彩！不妨一起製作彩虹蔬果海報，讓兒童進食不同顏色的蔬果時，貼上相應的彩色標籤以鼓勵攝取！

紅色	番茄、紅椒、西瓜、車厘子、士多啤梨。
	含脂溶性茄紅素，具抗氧化能力，以適量煮食油烹煮有助吸收。
橙黃色	紅蘿蔔、黃心番薯、哈密瓜。
	含 β - 胡蘿蔔素、葉黃素和玉米黃素，具抗氧化能力，有助防癌、維持眼睛、細胞及免疫系統健康。
綠色	西蘭花、菜心、羽衣甘藍、菠菜、蘆筍、牛油果、奇異果。
	含膳食纖維、類胡蘿蔔素、葉酸等重要營養素。
紫藍色	藍莓、布冧、茄子、紅菜頭。
	含水溶性花青素具抗氧化能力，有不少研究顯示或有助降低患上心血管疾病的風險。但較易流失於水分中，並受高溫破壞，建議以蒸及快炒方法盡量減少花青素流失。
啡白色	大蒜、洋葱、椰菜花、菇菌類。
	含蒜素，為天然抗微生物劑，有助抗菌。香蕉、薯仔更含有鉀質，有助保持血壓平衡。切碎蒜頭放置約 10 分鐘，有助提升蒜素含量，當遇熱時其抗氧化能力會隨煮食時間和溫度增加而降低。

肉、魚、蛋類及代替品

食物例子　：紅肉（豬、牛、羊）、家禽（雞、鴨）、海鮮（魚類、蝦、帶子）、非油炸豆製品（豆腐、鮮腐竹）、雞蛋、豆類（紅腰豆、鷹嘴豆、黃豆）和非油炸原味果仁（合桃、杏仁、芝麻、腰果）。

主要營養素：蛋白質、脂肪、維他命 B_{12}、礦物質（如鐵質）。

營養素	作用
蛋白質	促進人體生長發育和修補身體組織。 協助製造荷爾蒙，維持身體其他功能。
脂肪	分為飽和及不飽和脂肪（參閱 p.48）
維他命 B_{12}	增進食慾、協助製造紅血球、維持皮膚、眼睛和神經系統健康、幫助身體從蛋白質食物中釋放能量。
鐵質	肉類、家禽類、魚和海產含較容易吸收的血紅素鐵；植物性食物提供的鐵質較難被吸收，需要維他命 C 協助吸收。 製造紅血球，運送養分至身體各部分，預防缺鐵性貧血。

奶類及代替品

食物例子　：牛奶、芝士、乳酪、加鈣豆漿。

主要營養素：蛋白質、 鈣質、維他命 B_2。

營養素	作用
鈣質	鞏固骨骼和牙齒，預防骨質疏鬆。 幫助凝血、神經傳送和肌肉收縮。
維他命 B_2	協助碳水化合物、蛋白質和脂肪食物於體內分解並釋放能量。 維持口腔、眼睛、皮膚、毛髮和指甲健康。

油、鹽、糖類

食物例子：動物脂肪（牛油、忌廉）、植物油（橄欖油、芥花籽油、粟米油）、食用鹽、現成醬汁和調味料、糖分（砂糖、黑糖、蜜糖、糖漿）、零食（薯片、糖果、汽水）。

主要營養素：油（脂肪、脂溶性維他命）、鹽（鈉質）、糖（碳水化合物）。

營養素	作用
油分	提供能量，保護體內器官免受震盪。 分為飽和及不飽和脂肪（參閱 P.48）；攝取過量引致肥胖。 負責運送脂溶性維他命 A、D、E、K。
鹽分	提供鈉質，使細胞外液保持平衡；攝取過量增加患上高血壓風險。
糖分	提供能量和碳水化合物，於體內轉化為血糖；攝取過量會增加患蛀牙和肥胖的風險。

流質飲品

例子：清水、奶、清湯。

主要營養素：提供水分

營養素	作用
水分	補充身體因新陳代謝、出汗等所流失的水分。 維持體溫、傳送養分、氧氣和移走廢物。

＊不建議兒童飲用以下飲品：

含咖啡因飲品（刺激腦部中樞神經，導致心跳加速、手震，更會影響睡眠）：濃茶、咖啡、運動能量飲品。

高糖分或加入甜味劑飲品（增加患上蛀牙、肥胖和嗜甜的風險）：汽水、加糖果汁。

如何重「質」？

不同的營養素具有獨特的用處，對兒童成長十分重要。資訊發達，家長很容易從不同途徑接收營養資訊，但資訊孰是孰非？是否可信性高的資訊？的確，每位家長也愛錫子女，所以希望與家長分享幾個較常見的兒童營養問題，了解清楚再實踐吧！

1. 要預防發育遲緩？避免營養不良？想兒童健康成長？

攝取足夠的熱量和蛋白質有助預防營養不良，讓兒童健康成長。不同的食物均可轉化為熱量，提供能量使用，本港多以千卡（kcal）作單位。每 1 克的碳水化合物和蛋白質能提供約 4 千卡的熱量，而脂肪則提供 9 千卡的熱量。如兒童的能量和營養素攝取不足，會令身體發展遲緩，身高、骨骼和肌肉發展未如理想。首先，家長要了解兒童是否吃得足夠，應先重「量」（確保攝取足夠），再重「質」（仔細留意哪些營養素攝取不足）。

另外，足夠的蛋白質攝取有助促進兒童生長發育和修補身體組織，如能量攝取不足，蛋白質會分解以釋放能量供應身體，最終導致蛋白質能量營養不良。

2. 經常表現疲倦？面青青？指甲脆弱？胃口轉差？

有機會由於缺乏多於一種營養素而造成，但其中一種較常見的是缺乏足夠的鐵質攝取引致。聽到鐵質自然聯想到紅肉和貧血。鐵質有助製造紅血球，預防缺鐵性貧血，而紅血球更是血液中的元素，幫助運送氧分和養分到身體各器官使用。如身體缺乏鐵質，會減低紅血球製造，從而減低氧分和養分輸送，導致表現疲倦、臉色蒼白、指甲脆弱等情況出現。

有研究指出缺鐵性貧血人士因減少鐵質攝取降低飢餓素（Ghrelin）的水平，導致胃口不佳，進一步減少熱量和其他營養素的攝取。鐵質可從肉類、家禽類、魚和海產、深綠色蔬菜、全麥穀類、乾果中攝取；但過量攝取會引致便秘和肝臟衰竭的情況。

3. 兒童骨骼變軟、變形容易骨折？牙齒出現問題？

談到骨骼、牙齒，家長會聯想到鈣質；除了鈣質，維他命 D 的攝取亦不容忽視！鈣質有助強化骨骼和牙齒，除了從奶品類攝取外，深綠色或十字花科蔬菜（菜心、西蘭花）、芝麻、蝦米均能提供鈣質。維他命 D 則能協助身體吸收和使用鈣質，幫助牙齒和骨骼生長。一些高脂魚類（三文魚、鯖魚）、蛋黃、添加了維他命 D 的食物能提供維他命 D，但其含量較少，因此從陽光中的紫外線照射皮膚後在體內合成的維他命 D 才是主要來源。

兒童長期缺乏維他命 D 會引致佝僂病，因此應多作戶外體能活動（如一星期運動二至三次，每次約 15 分鐘），讓臉、手、腳短暫接觸陽光以增加攝取紫外線製造維他命 D 的機會。

4. 作為家長的辛酸，莫過於兒童容易生病……

兒童容易生病有機會由於抵抗力下降，以及免疫能力減低所致。攝取足夠的鋅質有助免疫系統正常運作，協助白血球正常發展及運作，如缺乏鋅質更會引致疲倦和脫髮等問題。一些低脂肉類及家禽、海產（蠔）、奶類製品、全穀物製品（糙米，麥包）、果仁及種子均含鋅質。另外，「皇牌維他命 A、C、E」有助調節免疫系統和維持抵抗力的功用。

維他命 A——加強細胞及黏膜的保護和再生，維持口腔、胃部、腸道及呼吸系統健康以防感染。橙黃色的蔬果如黃色番薯、紅蘿蔔、紅色燈籠椒、深綠色的蔬果如西蘭花、菠菜能提供維他命 A。

維他命 C——幫助製造膠原體，促進細胞（如抗體）的成長，協助鐵質吸收以加強抵抗力。每逢天氣轉換、流感高峰期，不少家長會購入維他命 C 補充品，但現時未有足夠的証據顯示大劑量維他命 C 有助預防傷風，如長期攝取過量更會造成腸胃不適，甚至增加患上腎石的風險，家長可從新鮮的蔬果攝取足夠的維他命 C。

維他命 E——具抗氧化功能，保護紅血球，維持細胞正常，保持皮膚及各組織健康。種子類及果仁、深綠色蔬菜、牛油果能提供維他命 E。

家長應緊記，沒有一種獨特的營養素能預防或治理所有疾病，均衡和多元化的飲食才能讓兒童攝取多種營養素。

為子女，選擇最好的！

選擇新鮮的食材當然重要，但隨着社會變化，一些預先包裝食品越來越普遍。要懂得選擇較健康的預先包裝食品，家長應閱讀包裝上的：1) 食用限期、2) 食材或成分表、3) 營養標籤，好好為子女的健康把關！

1. 食用限期

首先要看包裝是否完整無缺，留意該產品的食用期限。通常會發現食用期限日期前出現以下其中一種字眼：「此日期或之前食用」或「此日期前最佳」，到底兩者有何分別呢？

「此日期或之前食用」（Use by）的產品多為容易腐壞的食物，如鮮奶、乳酪、芝士、麵包、三文治等。這類食物只能存放較短的時間，過期後食用會對健康構成危險，因此如食物已過「此日期或之前食用」（Use by）的日期，即使表面上看似沒有問題亦應棄掉。

而「此日期前最佳」（Best before）則多列印於可存放較長時間的食物，如麵、餅乾、罐頭食品、零食等。這類食物若配合適當貯存，可預期其食物特質、味道、質感於該日期或之前食用屬最佳品質；但在該日期之後食用則未能保證其質素，因此應留意其品質再決定是否適合繼續食用。

2. 食材或成分表

成份：全麥麵粉(裂穀小麥，小麥)，小麥麵粉，菜籽油，脫脂奶粉，海鹽，百里香，酵母，大麥麥芽提取物，膨脹劑：碳酸鈉類。
本產品含有穀類(麩質)製品及奶類製品。可能含有微量榛子，杏仁(木本堅果)。
Ingredients: Whole meal flour (spelt, wheat), wheat flour, colza oil, skimmed milk powder, sea salt, thyme, yeast, barley malt extract, raising agent: sodium carbonates.
This product contains cereals containing gluten and milk products.
May contain traces of hazelnuts, almonds (tree nuts) products.

製作該食物的材料需列明於成分表內，並按照所佔的重量或體積，由大至小依次序列明。購買食物時發現油、鹽、糖分或同類材料列於較前的位置，則表明此產品含較高的脂肪、糖分和鹽分，不宜過量進食。

家長亦可從成分表留意產品是否額外加入糖分，例如果乾成分表發現「糖」、「蜜糖」等字眼，則表明產品添加了糖分。

如產品含有法例中列明的食物致敏物亦必須標示。另外，食物添加劑多列於較後的位置，因少量添加劑已能發揮其作用。

3. 營養標籤

於本港，營養標籤內需含有「1 ＋ 7」，分別為能量和 7 種營養素（碳水化合物、糖、蛋白質、總脂肪、飽和脂肪、反式脂肪和鈉質）。有些營養標籤列明了膳食纖維、膽固醇、鈣質等其他營養素的含量，由於不是法例上指定要求列明，食物製造商可自行決定是否需要列出。如包裝上列明了「高纖」、「高鈣」等營養聲稱的字眼，則需於營養標籤上列明該營養素的含量作參考。各營養素的含量一般可以「每 100 克 / 毫升」或「每食用分量」列明。

營養資料 Nutrition Information	每100毫升 Per 100mL
熱量/Energy	21 千卡/kcal
蛋白質/Protein	1.8 克/g
脂肪總量/Fat, total	1.2 克/g
- 飽和脂肪/Saturated fat	0.3 克/g
- 反式脂肪/Trans fat	0 克/g
膽固醇/ Cholesterol	0 毫克/mg
碳水化合物/Carbohydrates	0.7 克/g
- 糖/Sugars	0 克/g
鈉/Sodium	40 毫克/mg
鈣/Calcium	170 毫克/mg
維他命D/Vitamin D	1.0 微克/μg

以「每 100 克 / 毫升」標示：

列明如進食 100 克 / 毫升時，有多少能量和營養素可攝取到，較容易讓家長為同種類但不同生產商的食物作比較，快捷地找出較健康的選擇。

為「營養聲稱」作出定義，可參考下表以分別低脂、低糖、低鈉、高纖和高鈣的定義：

	每 100 克 不超過	每 100 毫升 不超過
總脂肪	3 克（低脂）	1.5 克（低脂）
糖	5 克（低糖）	
鈉	120 毫克（低鈉）	

	每 100 克 等於 / 超過	每 100 毫升 等於 / 超過
膳食纖維	6 克（高纖）	3 克（高纖）
鈣質	240 毫克（高鈣）	120 毫克（高鈣）

購買飲品時，如發現每 100 毫升含 5 克糖分，表示屬「低糖」較健康的選擇。不過應留意「低糖」不等於可以「隨意飲用」！當食用分量增加時，其糖含量亦會隨之增加。飲用一枝 500 毫升的低糖綠茶，已攝取 25 克糖分（即 5 茶匙糖），所以即使低脂、低糖和低鈉的食物亦要留意其食用分量。

以「每食用分量」標示：

這標示指一般人每次進食該食物時通常進食的分量，較容易知道食用該分量後可攝取的能量和營養素；不同的生產商會因該產品訂立不同的「建議進食分量」，某些牌子會減少其建議食用分量，令營養成分的數值看似較低、較健康，家長不應利用此標示與產品作比較，以免混餚而作出錯誤選擇。

Serving Size
食用份量：40 g 克
Servings Per Package
本包裝所含份數：5

		Per Serving 每份	Per 100g 每100克
Energy 熱量	Kcal/千卡	148	371
Protein 蛋白質	g/克	2.8	6.5
Total Fat 脂肪	g/克	0.4	1.0
Saturated Fat 飽和脂肪	g/克	0.2	0.4
Trans Fat 反式脂肪	g/克	0	0
Cholesterol 膽固醇	mg/毫克	0	0
Total Carbohydrates 總碳水化合物	g/克	34.0	85.0
Dietary Fibre 膳食纖維	g/克	0.4	1.0
Sugars 糖	g/克	3.6	9.0
Sodium 鈉	mg/毫克	184	460

舉例：牌子 A 的餅乾一盒有 4 包，每小包重 25 克為建議的「每食用分量」，含 10 克脂肪。牌子 B 的餅乾有 10 包，每小包重 10 克為建議的「每食用分量」，含 6 克脂肪。如一不小心利用「每食用分量」作比較，會誤以為牌子 B 屬較健康的選擇（因牌子 B 的 6 克脂肪較低），但這只是食品生產商製作出來的假象。將分量倍大至同等食用分量時（即兩者均為 25 克），牌子 B 的餅乾提供 15 克脂肪，比牌子 A 更高！因此，閱讀營養標籤時需小心，避免墮入陷阱！

營養標籤內的糖：

閱讀營養標籤要留心顯示的糖分，是整個產品所含的糖分，而非代表額外添加的糖分。購買一些低脂原味乳酪和含天然水果的食物時，當中的含糖量有機會頗高。由於低脂原味乳酪含有乳糖，水果當中含有果糖，均是糖分的一種，因此同被計算在營養標籤中的糖分含量，故不能單靠閱讀營養標籤得知額外添加糖分的含量。如希望知道額外添加糖分比例上的多少，則要依靠閱讀食物成分表，從而留意糖分排列的次序估計整個產品中糖分的所佔比例（含量愈多排在愈前位置）。

給家長的話

營養師聊天室

以上未有包括你的疑問？想得到更確實的應用方案？
不妨翻頁到食譜部分，一邊製作健康菜式，一邊繼續探討兒童常見的飲食問題吧！
的確，要為一餐半餐花時間準備食材、烹調、洗碗，然後一起溫習、閱讀圖書已經花費家長不少時間。明白要改變固有的習慣很難，但為了子女將來的健康着想，不如從今天開始與子女一同嘗試健康的新習慣、新食物，讓大家能培養健康的生活模式，預防疾病吧！

Chapter 2

入廚樂，親子互動煮食！

2.1 親子煮食好處多

家長與兒童一起製作食物，不但能促進親子關係，更可：

- 訓練兒童的耐性。
- 增強自理能力和學習多種生活技能。
- 發揮創意和加強自信心。
- 增強不同語言的閱讀理解和與人溝通能力。
- 訓練數學及換算理念。
- 減少使用屏幕的時間。
- 讓兒童接觸不同的食材，增加對食物的興趣，預防和解決偏食問題。
- 訓練兒童的手部操作和協調，例如：
  - 1-2 歲的兒童：一般可用雙手拿起食物或包裝食品，並將它們疊高。
  - 2 歲以上的兒童：可試用膠刀切較軟身的水果（如香蕉），訓練手指操作。
  - 3 歲以上的兒童：可透過搓壓麵糰及製作點心，提升眼睛和雙手的協調，以及手指的靈活性，更提升小朋友的自理能力和自信心。

小提示

製作食物時，家長不妨趁機教導兒童珍惜食物及了解食物的營養，從而培養健康飲食習慣。提提家長，應按兒童的能力耐心教導，讓他們在愉快的環境下練習，切忌操之過急或過早催谷兒童。兒童在製作過程中有錯處，算是一個學習及製造回憶的機會，不用過分認真，多鼓勵、少責罵；當然要緊記安全至上。

2.2 兒童可參與製作食物的步驟

家長可根據兒童的年紀、入廚經驗和能力，自行評估和決定兒童應如何參與食物製作的過程，步驟如下：

- 設計餐單——多與兒童翻閱食譜書籍，讓小朋友選擇並參與餐單設計 。
- 選購食材——與兒童一起到超市或街市選購食材，讓兒童擴闊視野，接觸更多食物種類，學習如何精明地選購食材（食譜內的食材備有方格作清單選購，讓兒童記錄該食材是否已購買或預備）。
- 預備材料及製作食品。
- 上菜和擺盤裝飾。
- 清潔和儲存食物。

年齡較小的兒童可參與較簡單的工作，如：清洗食材；量度分量；將食材倒入大碗；將食物壓成蓉；塗麵包；擺盤裝飾等 。
年齡較大的兒童可試試以下任務，如：將食物去皮、切件；處理肉類（醃肉和搓肉丸）；煎雞蛋或熱香餅；準備及清洗材料及餐具，以至自行負責製作菜式等。

2.3 食物安全 5 要點，家長要留意！

家長與兒童製作食物時，不能忽略的五大食物安全要點：

1. CHOOSE 精明選購食物

鼓勵到有信譽、衛生的商舖選購食物，同時留意以下幾點：

- 新鮮食材：蔬果表面不應有破損或瘀傷；雞蛋不應有裂縫或滲漏。
- 包裝食物：包裝不應有破損。
- 罐裝食物：罐身並無膨脹或凹陷。
- 玻璃瓶食物：玻璃瓶不應有裂縫；檢查瓶蓋不應鬆動或有開啟的痕跡。
- 留意食用限期，不要選購已經過期的食物（詳細指引可參閱 p.20）
- 熟食及較容易腐壞的食物應於兩小時內存放雪櫃內。

2. CLEAN 保持乾淨

- 處理食物、進食前、撫摸寵物及上洗手間後，緊記清洗雙手，洗手 5 部曲如下：
  - 先用清水弄濕雙手，加入適量洗手液；
  - 搓雙手至起泡泡；
  - 起泡泡後，徹底清潔雙手，可邊洗手邊唱生日歌（至少 2 次，約 20 秒）；
  - 用清水沖走泡泡；
  - 最後用乾淨的毛巾抹乾雙手。
- 保持工作枱面整潔，餐具和砧板要確保乾淨才使用；製作食物後用熱水及清潔劑徹底清洗廚具及工作枱。
- 處理或食用蔬菜、水果前必須用水清洗乾淨。

3. SEPARATE 生熟食物要分開

- 購物、存放及處理食物時，緊記生熟食物必須分開。
- 避免生肉、海鮮和雞蛋的細菌傳播到其他食物。
- 為免交叉污染，建議預備不同的砧板方便分類，更可以不同顏色區分使用：蔬菜、水果；生肉；熟食。

4. COOK 徹底煮熟

- 不建議單靠食物的表面顏色和質感來判斷是否徹底煮熟，最安全的方法是用食物溫度計量度食物的中央位置溫度。
- 食物離開熱源前，確保食物已加熱足夠，食物中央（遠離骨頭和多脂肪位置）的溫度達至最低建議溫度，如下表：

食物	建議溫度
牛肉、豬肉和羊肉	145 ℉（62.8℃），最少靜置 3 分鐘
肉碎	160 ℉（71.1℃）
雞肉碎	165 ℉（73.9℃）
雞肉	165 ℉（73.9℃）
雞蛋	160 ℉（71.1℃）
魚、貝殼類海鮮	145 ℉（62.8℃）
剩餸	165 ℉（73.9℃）

不少兒童都喜愛吃香脆的食物，薯片、薯條、炸雞翼等均是他們的最愛。作為家長，當然知道這些經高溫油炸的食物不健康，多吃會引致肥胖和患上心血管疾病的風險。

作為一位「知慳識儉」的家長，更不會在家中利用一整鍋油來煎炸食物吧！有見及此，近年不少人士熱捧以氣炸鍋製作食物，認為鍋子細細並只利用少量油可將食物變得香脆可口屬減肥恩物。的確，以氣炸鍋製作食物相比傳統油炸能減少油的使用量，以致整體熱量相對降低，有利控制體重。

相比一般焗爐，氣炸鍋體積較輕便，其預熱功能及烤焗時間亦較快，難怪深得大家歡心；但要留意不少經高溫製作的食物會使其主要營養成分如：碳水化合物、脂肪和蛋白質產生變化並產生致癌物。以高溫烹煮含豐富碳水化合物和較少蛋白質的植物性食物薯條為例，會產生丙烯酰胺（Acrylamide），當烹調溫度愈高時間愈長，產生致癌物質愈多，因此不建議經常吃燒烤、烘焗或煎炸食物（尤其避免燒焦）。想減少食物當中的致癌物質，不宜以過高的溫度烹煮過長時間。當製作煎炸食物時，不論以氣炸、烘焙、烤烘或燒烤形式製作，應把食物煮至呈金黃色或淺黃色即可。

5. CHILL 妥善存放食物

- 較容易變壞的食物不應放在室溫超過 2 小時。
- 細菌適合生長的溫度範圍稱為「危險溫度範圍」，即攝氏 4 至 60 度之間。
- 如有剩餸應盡快放入有蓋容器內，再存入雪櫃冷藏，並於 3-4 日內食用。
- 安全解凍食物很容易，可預先計劃第二天要使用的食材，有充裕時間以較安全的方法解凍：
  - 前一天晚上將解凍的食物用保鮮紙包好，放入雪櫃解凍。
  - 已煮熟或即食的食物應放在雪櫃上層；未經煮熟或非即食食物應放雪櫃下層。
  - 食物解凍後應於兩天內處理及食用，並不建議再次冷藏以防細菌性食物中毒。
  - 利用微波爐或流動冷水解凍也是建議的方法。

Chapter 3

零失敗，學懂煮食基本功！

3.1 認識不同的烹飪工具

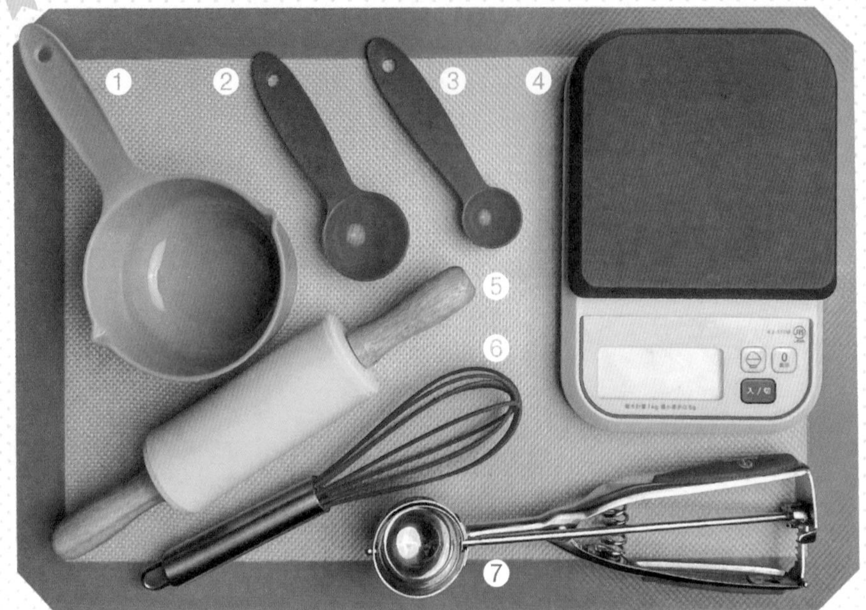

❶ 量杯（一杯分量）

在沒有電子磅的情況下，用作量度一杯的分量，如麵粉、液體等；但要留意不同食材的重量會有異，如一杯液體約 240 毫升；一杯麵粉則約 100 克。

❷ 量匙（一湯匙分量）

在沒有電子磅的情況下用作量度，一般約為 15 毫升。

❸ 量匙（一茶匙分量）

在沒有電子磅的情況下用作量度，一般約為 5 毫升。煮食時建議用量匙下油，以控制煮食油的使用量。

❹ 電子磅

準確地量度材料和食材的重量，使用時緊記先將電子磅的重量歸「零」，再進行量度。

❺ 擀麵棒

製作麵包、薄餅、麵條時，有助將麵糰壓平。

❻ 拂打器

例如用於拂打雞蛋製作茶碗蒸，或製作甜點時拂打低脂忌廉，充分加入空氣，讓低脂忌廉變得軟滑。

❼ 圓球勺

製作鮮果雪葩時使用，能製作精美的球狀。製作薯蓉或南瓜蓉亦可用作上菜擺盤，以增加兒童對菜式的興趣。

⑧ 剪刀

將食物剪碎或去除較細小的脂肪部位，應由家長使用並放置於兒童不能觸及的地方。

⑨ 膠刀（兒童用）

兒童準備材料時可使用膠刀，協助家長切開較軟的食材，如香蕉、麵包等。

⑩ 刀（家長用）

較鋒利的刀應由家長使用，並放置於兒童不能觸及的地方。

⑪ 膠刮刀

將食物混合和攪拌，耐熱度較高的膠刮刀也適合用作快炒食物（使用前應留意包裝說明）。

⑫ 膠掃

烘焗前為食物塗上蛋漿，營造金黃色的效果；或將食用油塗在易潔鑊上，以控制用油量。

⑬ 膠勺

常用作混合、攪拌和壓蓉食物。

⑭ 抹刀

將醬汁塗抹在碟上作裝飾，增加兒童對菜式的興趣。

3.2 餐具選配，了解進食習慣

可提供不同的餐具，讓兒童選擇合適的餐具進食。利用分隔碟子有助了解兒童進食穀物類、蔬菜和肉類的分量，更可分隔醬汁，減少「撈汁食飯」的機會，以進一步降低額外糖分和鹽分的攝取。

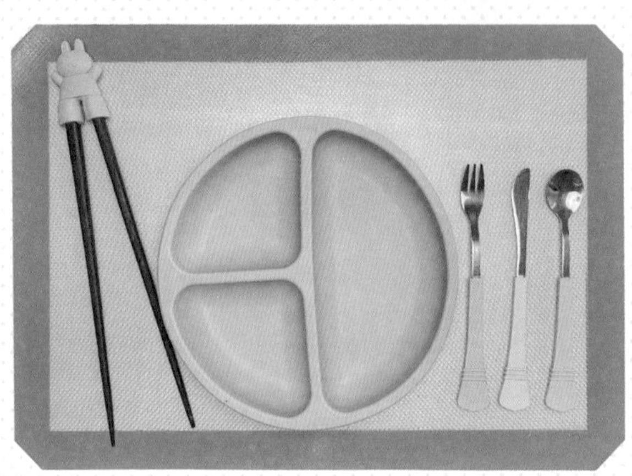

3.3 學會各款食材切法，提升廚藝

學懂並利用不同的食物切法，能將食材完美地運用於不同種類的菜式中。不妨學習以下的切法，助你的廚藝逐步升級！因應兒童的咀嚼程度接觸不同食材的質地，讓兒童容易嘗試新食材。

雞扒

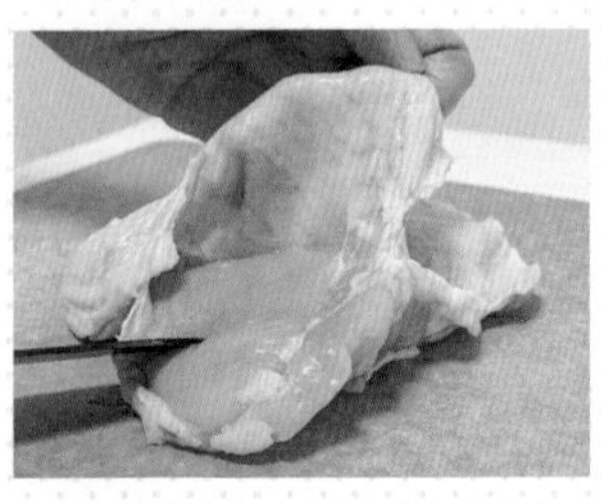

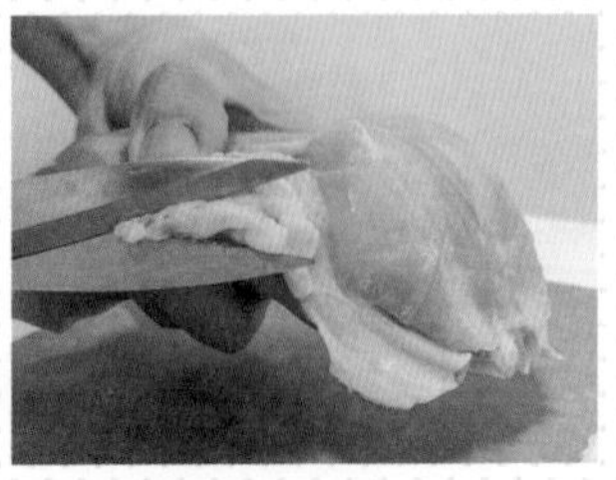

家長先用刀將大部分動物皮層去掉，再用剪刀清除較少部分或較難去掉的脂肪，以減少飽和脂肪的攝取。

肉類

✔

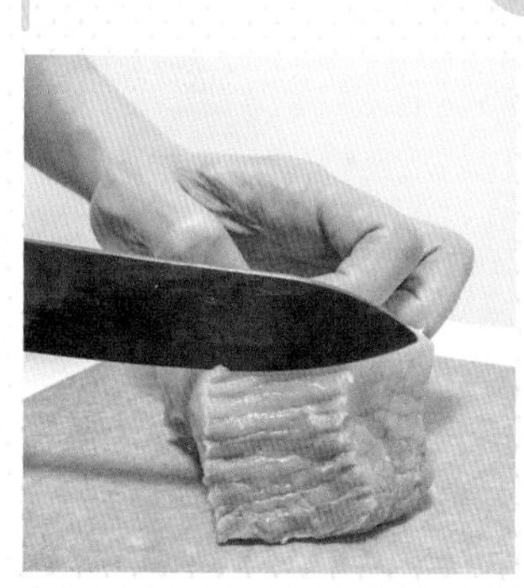

逆紋切肉使肉質變得軟滑，較易咀嚼。

X

順着肉的紋理切開，使肉質變硬，較難咀嚼。

肉類的不同切法，運用於不同的烹調方法，符合菜式要求之餘，也令兒童增加不同的口感。

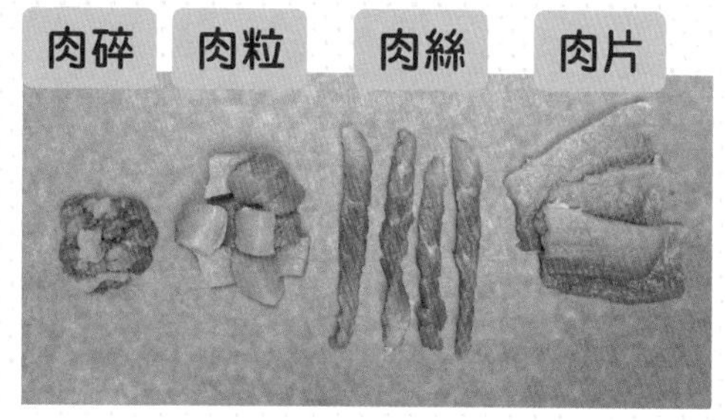

瓜類

使用「滾筒型切法」使瓜類形成不規則形狀，增加表面的面積，有助加熱及容易煮熟。

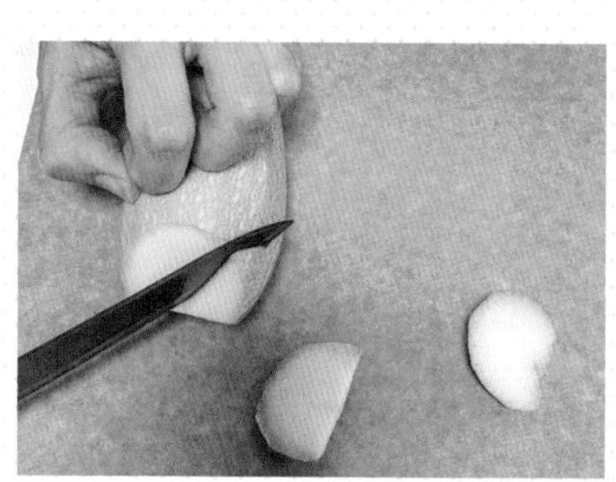

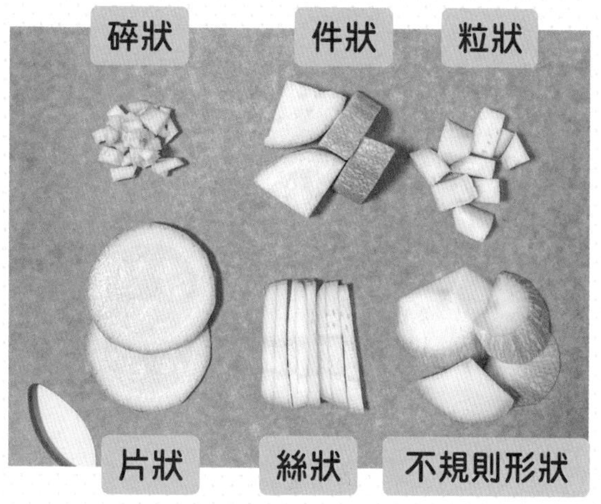

燈籠椒

將燈籠椒底部向上，以刀將每邊 1/3 部分燈籠椒由上而下切開，可輕易地將籽去掉。

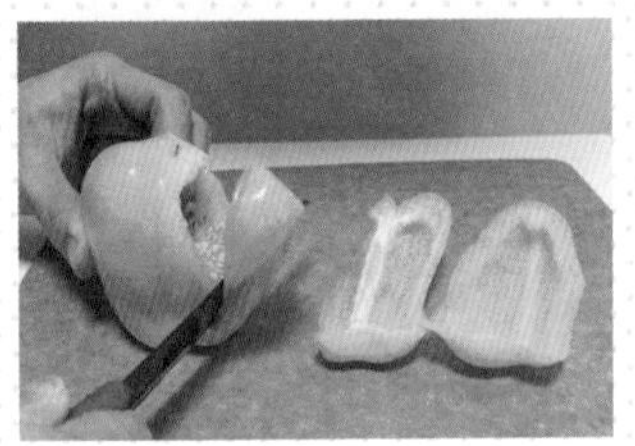

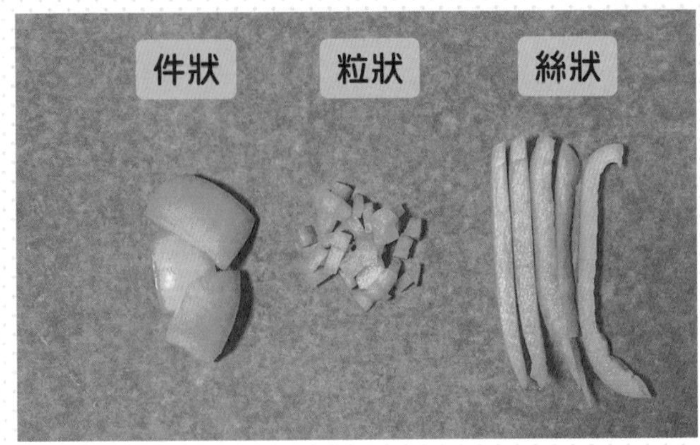

芒果

在果肉切成小格子，翻開後以湯匙取出果肉。

牛油果

將牛油果切十字，可輕易地切成 1/4 件果肉，拔出果核，輕鬆地逐件如香蕉般去皮。

洋蔥

將洋蔥兩端去掉，形成平面方便垂直切半，去掉外皮。半個洋蔥平放砧板，垂直切成條狀（保留一端不切斷），轉 90 度，切成粒狀即可。

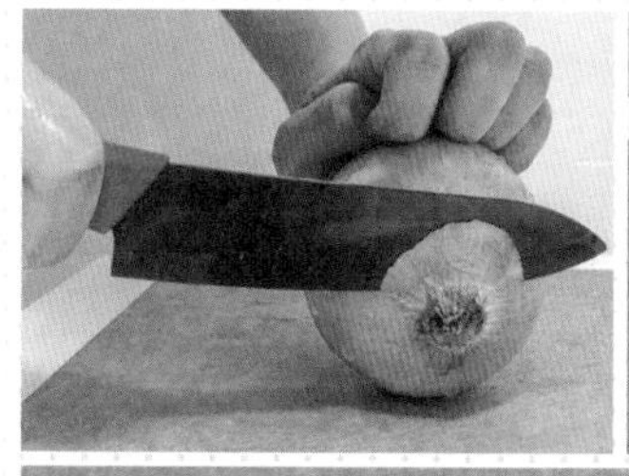

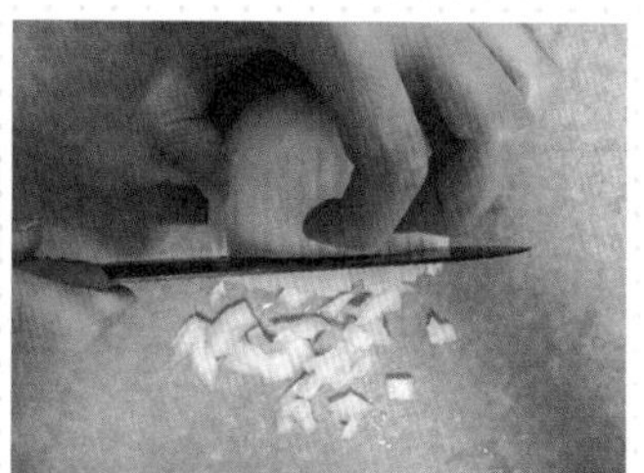

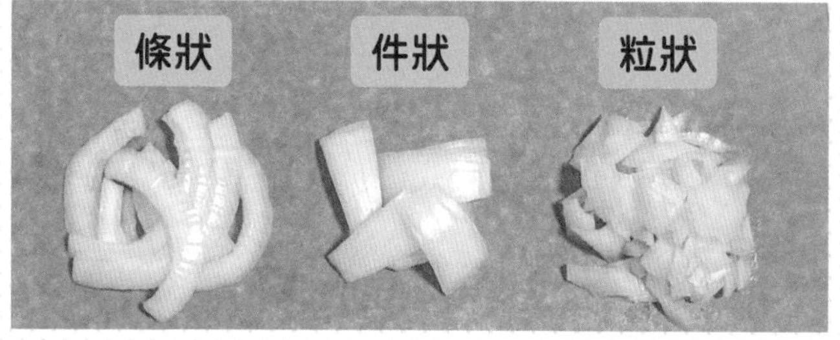

蒜頭、薑

將蒜頭兩端切掉，以刀按壓蒜頭方便去皮（留意刀刃向外）。

薑洗淨，用茶匙由下而上刮掉薑皮。

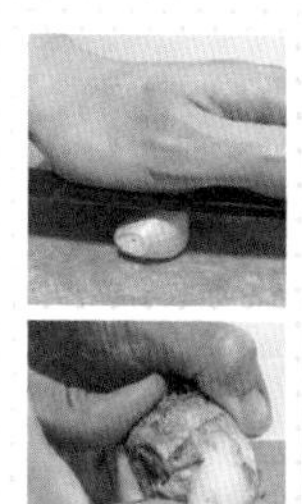

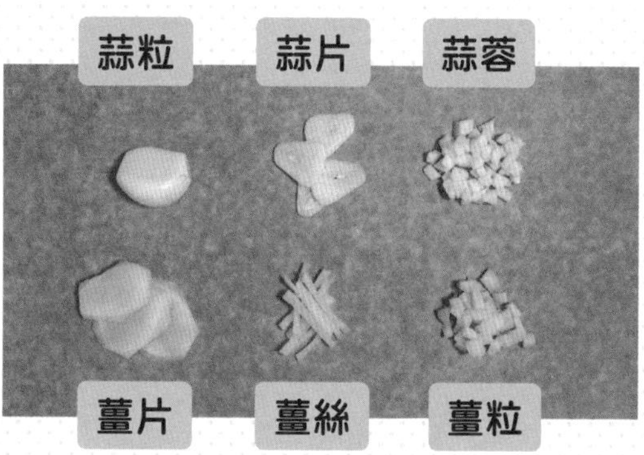

小提示

運用天然調味料如洋蔥、檸檬、香草、薑、葱、蒜頭等，能增加菜式的天然風味，從而減少糖分和鹽分的使用量。

兒童玩樂區，認識有營食材！

家長及兒童認識了不同的營養知識後，先在玩樂區熱熱身！兒童可發揮創意，參考以下圖畫進行遊戲，加深學習食材小知識，再跟父母享受入廚之樂趣！

Chapter 4

動動手，變出得意造型！

| 6 人分量 |

小熊香蕉蛋糕

Teddy Bear Steamed Banana Cupcakes

每份 (per serving)
* 不包括裝飾 (excluding the toppings)

熱量 (Energy)	碳水化合物 (Carbohydrate)	蛋白質 (Protein)	脂肪 (Fat)	膳食纖維 (Dietary Fibre)
103	17	3	3	1
千卡 (kcal)	克 (g)	克 (g)	克 (g)	克 (g)

材料

- ☐ 自發粉 75 克
- ☐ 雞蛋 1 隻
- ☐ 蔗糖 1 湯匙
- ☐ 植物油 1 湯匙
- ☐ 低脂鮮奶 25 克
- ☐ 全熟香蕉 1 隻（中型，壓成蓉）

裝飾

- ☐ 提子乾（無添加糖）、白朱古力、南瓜籽、朱古力醬或花生醬

做法

1. 於蒸鍋加入熱水，備用。
2. 雞蛋放入大碗內，攪拌至起泡沫，加入植物油、蔗糖和鮮奶，攪拌約 5 分鐘至混合，見顏色變淺和滑溜，拌入香蕉蓉。
3. 輕輕地將已過篩的自發粉分 3-4 次加入麵糊，拌勻，倒入 6 個紙杯模至九成滿。
4. 蛋糕放入蒸鍋，蒸約 15 分鐘至熟透，取出。
5. 最後以提子乾、白朱古力、南瓜籽和朱古力醬或花生醬裝飾，即成。

小助手參與步驟

營養小貼士

- 食譜加入香蕉不但能增加膳食纖維和多種營養素，其口感和甜味更可以取代部分油脂和添加糖，讓整個蛋糕更健康！
- 食譜利用植物油代替傳統蛋糕使用的牛油，能減少不健康的飽和脂肪攝取。
- 選購提子乾和種籽類時，建議多選無添加糖和鹽的種類。
- 除了提子乾和南瓜籽外，可與兒童一起討論其他水果乾和果仁作材料，增加他們對食物的興趣。

| 8 人分量 |

蝴蝶結 / 領結麵包

Butterfly / Bowtie Bun

每份 (per serving)

熱量 (Energy)	碳水化合物 (Carbohydrate)	蛋白質 (Protein)	脂肪 (Fat)	膳食纖維 (Dietary Fibre)
167	27	5	4	1
千卡 (kcal)	克 (g)	克 (g)	克 (g)	克 (g)

材料

- ☐ 高筋麵粉 250 克
- ☐ 糖 25 克
- ☐ 鹽 2 克
- ☐ 酵母粉 3 克
- ☐ 雞蛋 1 隻
- ☐ 低脂鮮奶 100 克
- ☐ 芥花籽油 25 克
- ☐ 高筋麵粉少許（灑面用）

營養小貼士

- 麵包加入雞蛋和低脂奶製作，增加兒童的蛋白質攝取。
- 麵包造型可愛，大小適中，能吸引兒童日常進食。
- 吃一個麵包已提供約半碗飯的碳水化合物，可配合一份低脂奶類食物或飲品（如低脂芝士兩片或低脂奶一杯），成為健康早餐之選擇。
- 可考慮逐步將 1/3 的高筋麵粉換成高筋麥麵粉或燕麥粉，有助進一步增加膳食纖維含量。

做法

1. 於大碗內加入高筋麵粉、糖、鹽和酵母粉拌勻（加入時留意酵母粉與糖和鹽分開放置，以免被破壞）。
2. 加入雞蛋和低脂鮮奶後，將麵糰混合至不黏手，取出，放在桌面搓揉至有筋性和表面光滑（搓揉方法：先按着麵糰下方，用手心將麵糰向上推再捲回，重複此步驟）。
3. 取出一小部分麵糰，用手指慢慢拉伸至有破洞。
4. 如破洞邊緣平滑，可逐少加入芥花籽油，再搓揉讓麵糰慢慢吸收油分（剛加入油時，麵糰會變得鬆散黏手，不用擔心繼續搓揉）。
5. 將麵糰搓成型，包成團狀放於碗內以保鮮紙蓋好，發酵 1 小時或至原有麵糰之兩倍大。
6. 發酵好的麵糰放桌面用手按壓排氣，分成 8 份（每份約 50 克），每小份摺好滾圓。
7. 蓋上保鮮紙，讓麵糰放鬆 20 分鐘，以擀麵棒滾平成圓形，切成右圖示的兩款形狀。
8. 將摺疊好的蝴蝶結／領結麵糰放在已鋪上牛油紙的焗盤，蓋上保鮮紙進行第二次發酵（約 40 分鐘）。
9. 預熱焗爐 180℃，於麵糰灑上少許高筋麵粉，放入焗爐約 15 分鐘即成。

小助手參與步驟（步驟 2、4、6、7）

蝴蝶結造型

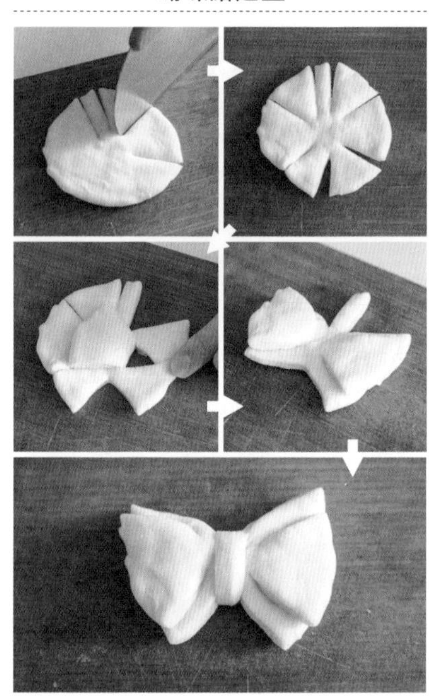

領結造型

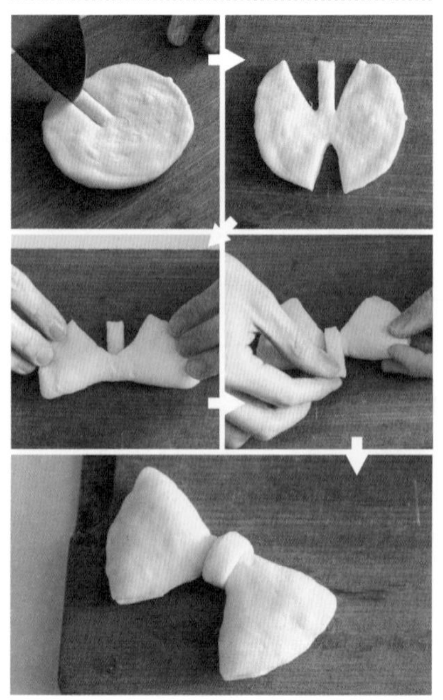

| 8 人分量 |

甜栗南瓜包

Pumpkin Bread with Chestnut Filling

每份 (per serving)

熱量 (Energy)	碳水化合物 (Carbohydrate)	蛋白質 (Protein)	脂肪 (Fat)	膳食纖維 (Dietary Fibre)
176	31	4	4	1
千卡 (kcal)	克 (g)	克 (g)	克 (g)	克 (g)

材料

- ☐ 南瓜 100 克
- ☐ 高筋麵粉 210 克
- ☐ 低脂鮮奶 50 克
- ☐ 糖 25 克
- ☐ 鹽 2 克
- ☐ 酵母粉 3 克
- ☐ 芥花籽油 20 克
- ☐ 即食栗子 8 顆
- ☐ 南瓜籽 8 粒

做法

❶ 南瓜去皮、去籽，切小粒，以中火隔水蒸 15 分鐘至軟身，壓蓉，放涼備用。

❷ 大碗內放入高筋麵粉、南瓜蓉、糖、鹽和酵母粉拌勻。逐少加入低脂鮮奶搓成麵糰（每次蒸熟的南瓜含水分不一，先留意麵糰的質地宜逐少加入鮮奶）。

❸ 將麵糰混合至不黏手及碗，放桌面搓揉至有筋性和呈光滑表面。（搓揉方法：先按着麵糰下方，用手心將麵糰向上推再捲回，重複此步驟。）

❹ 取出一小部分麵糰，用手指慢慢延伸至有破洞。如破洞邊緣平滑，可逐少加入芥花籽油再次搓揉，讓麵糰慢慢吸收油分（剛加入油分時，麵糰變得鬆散黏手，繼續搓揉即可）。

❺ 將麵糰搓成型，包成團狀放於碗內，以保鮮紙蓋好發酵 1 小時或至原有麵糰體積的兩倍。

❻ 發酵好的麵糰放桌面，用手按壓排氣，分成 8 份（每份約 50 克），每小份搯好滾圓。蓋上保鮮紙，讓麵糰放鬆 20 分鐘，以擀麵棒滾壓成圓形，放入即食栗子包好。

❼ 用牙籤或刮刀沾上少許高筋麵粉，在麵糰壓出「米字形」南瓜模樣，在頂端加上南瓜籽裝飾。

❽ 將南瓜形麵糰放在已鋪上牛油紙的焗盤，蓋上保鮮紙讓麵糰作第二次發酵約 40 分鐘。

❾ 預熱焗爐 180℃，焗約 15 分鐘即成。

小助手參與步驟

南瓜造型

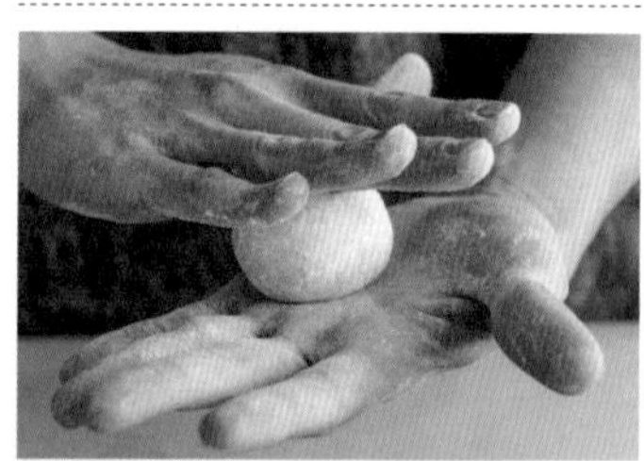

- 以南瓜製成麵包能增加膳食纖維攝取並提供類胡蘿蔔素（於體內轉化成維他命 A），有助眼睛健康和預防夜盲症。
- 類胡蘿蔔素為脂溶性維他命，過量攝取會留於體內，令皮膚呈橙黃色；只要停吃橙黃色食物和綠葉蔬菜一段日子即可。

營養師聊天室

兒童 poo-poo 不暢順，怎麼辦？

無論是兒童、成年人或長者均有機會遇到便秘的煩惱。由於排便習慣人人不同，因此正常的排便次數可以由每天多於一次或相隔兩天才一次不等。除了次數，家長應留意兒童大便的質地，如質地太硬且排便困難或需費力排便，有機會是便秘問題。

飲食中缺乏膳食纖維和水分會令大便變硬，難以排出。過量進食鈣片或含鈣的營養補充品也會導致便秘問題。家長應鼓勵兒童多進食新鮮水果和蔬菜，以攝取足夠的膳食纖維。建議兒童每天的膳食纖維攝取量為其年齡加 5 克（例如：一位 6 歲兒童每天膳食纖維攝取量大概是 6 + 5 = 11 克）。

選擇高纖維的穀物類食物，如紅米、糙米混合白米、全穀麥麵包等均有助增加攝取膳食纖維，建議以循序漸進的方式增加，以免導致肚脹、排氣、肚瀉等問題。家長不妨多提醒兒童喝清水，攝取適量流質飲品有助腸臟蠕動，再配合適量運動和養成定時排便的習慣，相信便秘的問題可慢慢解決。

食物	糙米（熟）1 杯	藜麥（熟）1 杯	粟米（熟）1 杯	紅蘿蔔（熟）1 杯	西蘭花（熟）1 杯	番石榴 1 杯	紅莓 1 杯	香蕉（中型）1 條	鷹嘴豆（熟）1/2 杯	青豆（熟）1/2 杯	杏仁 1 安士
膳食纖維	3.2 克	5.2 克	4.0 克	4.8 克	5.2 克	8.9 克	8.0 克	3.2 克	6.3 克	4.4 克	3.5 克

香滑果醬，大人、兒童都喜歡！
HAVE FUN
特色輕食

| 4 人分量 |

日常牛油果醬 Your Everyday Avocado Spread

每份 (per serving)

熱量 (Energy)	碳水化合物 (Carbohydrate)	蛋白質 (Protein)	脂肪 (Fat)	膳食纖維 (Dietary Fibre)
69	4	1	6	3
千卡 (kcal)	克 (g)	克 (g)	克 (g)	克 (g)

材料

- ☐ 牛油果 1 個（中型）
- ☐ 番茄粒 2 湯匙
- ☐ 洋葱粒 2 湯匙
- ☐ 蒜蓉 1 茶匙
- ☐ 橄欖油 1 茶匙
- ☐ 檸檬 1/4 個（榨汁）
- ☐ 鹽少許
- ☐ 黑椒碎少許

做法

❶ 鍋內加入油，放入番茄粒、洋葱粒和蒜蓉，以中小火炒香，放在碗內待涼。

❷ 牛油果切開，去核，用匙羹取出果肉。

❸ 所有材料放在大碗混合，並壓碎牛油果，加入檸檬汁、鹽和黑椒碎拌勻，即可享用。

小助手參與步驟

營養小貼士

- 牛油果含豐富膳食纖維、鉀質及維他命 E 等抗氧化物，同時富含不飽和脂肪酸，有利心臟健康。
- 每100克（約半個）牛油果已提供約15克脂肪（3茶匙油），雖屬於健康的不飽和脂肪，但對需控制體重的人士不宜過量攝取。
- 牛油果質地軟滑，配上洋葱粒和番茄粒，能豐富牛油果醬的口感，更能增加膳食纖維的攝取。
- 檸檬含有豐富維他命 C，更有助預防牛油果因氧化而變黑。

好油壞油怎樣分？

人體需要適量脂肪維持身體正常運作，功能包括：適當提供能量、維持體溫、保護內臟器官及輔助傳送和吸收脂溶性維他命 A、D、E 及 K 等。脂肪一般可分為以下類別：

	飽和脂肪	反式脂肪	單元不飽和脂肪	多元不飽和脂肪
室溫下狀態	固體	半固體或固體	液體	液體
對身體影響	增加總膽固醇及壞膽固醇水平。 增加患上心臟病風險。	增加壞膽固醇水平。 降低好膽固醇水平。 增加患上心臟病風險。	降低總膽固醇及壞膽固醇水平。 減低患上心臟病風險。	降低總膽固醇及壞膽固醇水平。 減低患上心臟病風險。 加快清理血液中的壞膽固醇，降低壞膽固醇水平。 攝取過量影響好膽固醇的分量。
食物來源	動物性脂肪：肉類、牛油、豬油、忌廉。 植物性脂肪：椰子油、棕櫚油。	部分人造牛油、起酥油、酥皮、批撻、經高溫烘焗或油炸的高脂食物。	橄欖油、芥花籽油、牛油果、非油炸原味果仁。	人體不能自行製造奧米加 3（Omega-3）和奧米加 6（Omega-6）脂肪酸，必需透過食物攝取。 粟米油、大豆油、三文魚、核桃、葵花籽。

小提示：

建議選擇包裝食物時，留意成分表有沒有氫化植物油、部分氫化植物油、起酥油等字眼，並留意營養標籤中飽和及反式脂肪的含量是否過高。煮食時多採用不飽和脂肪的植物油，並以無添加果仁醬或牛油果代替牛油塗麵包，以不飽和脂肪逐步代替飽和脂肪。

| 2 人分量 |

羽衣甘藍脆脆 Curly Kale Crunch

每份 (per serving)

熱量 (Energy)	碳水化合物 (Carbohydrate)	蛋白質 (Protein)	脂肪 (Fat)	膳食纖維 (Dietary Fibre)
37	2	1	3	1
千卡 (kcal)	克 (g)	克 (g)	克 (g)	克 (g)

材料

- □ 羽衣甘藍 2 杯（取葉片）
- □ 橄欖油 2 茶匙

調味料

- □ 鹽 1/8-1/4 茶匙
- □ 青檸汁 1/4 個

做法

1. 把焗爐預熱至 175℃。
2. 所有材料放入大碗，攪拌至全部羽衣甘藍沾有橄欖油。
3. 平均地鋪在上烤盤，焗約 10 分鐘至變脆，按個人喜好加入適量鹽及青檸汁即成。

小助手參與步驟

營養小貼士

- 羽衣甘藍一直被譽為超級食物，全因其營養豐富，含有多種抗氧化物如維他命 A、維他命 C、維他命 K、葉黃素、玉米黃素及鈣質、鉀質等。
- 每 100 克羽衣甘藍只含 35 千卡，並提供約 4 克膳食纖維，屬高纖低熱量食品。
- 適量進食羽衣甘藍對體重管理有利，更能預防癌症、維持眼睛和整體健康。
- 羽衣甘藍脆脆口感鬆脆，可代替不健康的薯片以減少油分和鹽分的攝取。

飲食與眼睛健康有關嗎？

與眼睛健康有關的營養素除了花青素（Anthocyanins）外，還有葉黃素（Lutein）、玉米黃素（Zeaxanthin）、維他命 A、維他命 C、維他命 E、鋅質（Zinc）和奧米加 3 脂肪酸等。以下有幾類護眼食物的例子：

- 葉黃素及玉米黃素有助阻隔藍光及紫外線對眼睛的傷害，深綠色蔬菜是重要的食物來源，羽衣甘藍、芥蘭、菠菜、西蘭花等都是不錯的選擇，配合適量油分進食有助吸收！
- 種籽果仁如杏仁、葵花籽、花生等含有豐富維他命 E 及健康脂肪酸，適量進食有助維持心臟健康；但需留意熱量偏高，14 粒原味杏仁或 17 粒原味花生肉已提供約 100 千卡。另外應選擇非油炸原味果仁以進一步減低脂肪和鹽分的攝取。
- 三文魚含有豐富奧米加 3 脂肪酸，有助保護眼睛及紓緩眼乾症狀。其他含豐富奧米加 3 脂肪酸的魚類包括沙甸魚、鯖魚、黃花魚等。
- 雞蛋烹調簡單，也可配搭不同的食材，能提供優質蛋白質，是一種方便百搭的食材，當中含有鋅質、葉黃素及玉米黃素等營養素。其他含豐富鋅質的食材包括蠔、牛肉、海鮮、南瓜籽和豆類等。
- 莓類是很好的食材，除藍莓外，士多啤梨含有豐富維他命 C 及抗氧化物！其他含豐富維他命 C 的蔬果包括紅甜椒、西蘭花、檸檬、奇異果和橙等。

最後提提你，建議適量補充水分，有助避免眼乾問題。使用電子產品時，每 20 分鐘讓眼睛休息至少 20 秒！

| 2 人分量 |

西多士方塊 Hong Kong Style French Toast

每份 (per serving)

熱量 (Energy)	碳水化合物 (Carbohydrate)	蛋白質 (Protein)	脂肪 (Fat)	膳食纖維 (Dietary Fibre)
194	27	8	7	4
千卡 (kcal)	克 (g)	克 (g)	克 (g)	克 (g)

材料

- ☐ 方包 2 片（大，全麥麵包較佳）
- ☐ 無添加花生醬 2 茶匙 -1 湯匙
- ☐ 香蕉 1/2 隻（全熟，壓蓉）
- ☐ 雞蛋 1 隻
- ☐ 低脂鮮奶 2 湯匙
- ☐ 植物油 1 茶匙
- ☐ 藍莓及士多啤梨 1/2 杯
- ☐ 楓糖漿 / 蜂蜜（按需要）

做法

1. 於大碗內加入雞蛋和低脂鮮奶，攪勻備用。
2. 方包去邊，塗上花生醬及香蕉蓉，將兩片方包夾好，切成 4 小塊或不同形狀，放進蛋液。
3. 預熱平底鍋，平均塗上少量油，放入沾滿蛋液的三文治，用中火慢慢煎至兩邊金黃色。
4. 將西多士放在碟上，加上藍莓及士多啤梨，淋上少量糖漿即成。

小助手參與步驟

營養小貼士

- 比一般油炸的西多士，此食譜以少量油分製作西多士，大大減低脂肪的攝取。
- 用花生醬和香蕉代替忌廉或煉奶餡料，讓西多士營養更豐富。
- 用清甜的藍莓、士多啤梨代替部分或全部糖漿，營養更豐富，顏色更鮮艷。

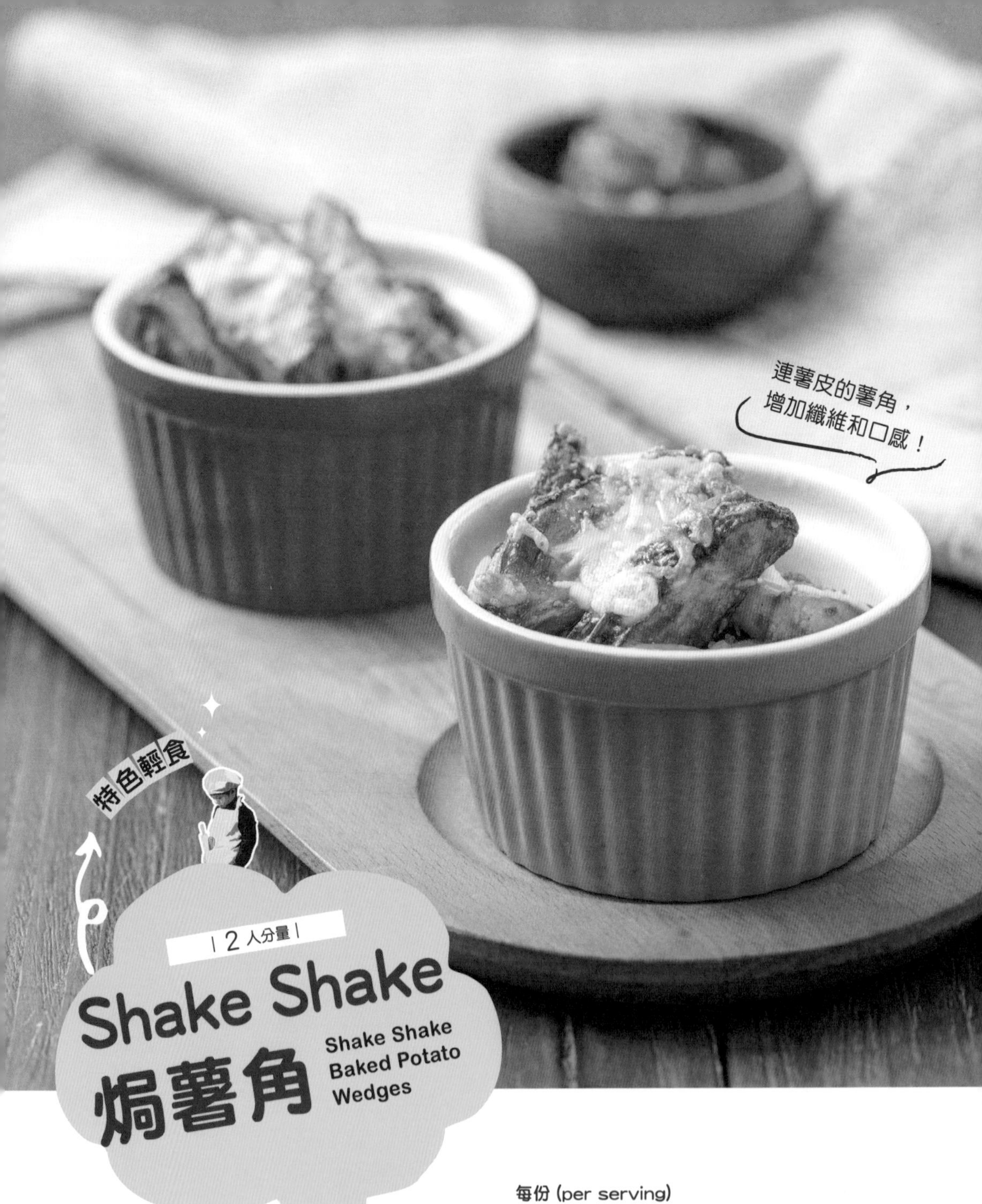

| 2 人分量 |

Shake Shake 焗薯角

Shake Shake Baked Potato Wedges

每份 (per serving)

* 不包括醬汁 (excluding the creamy yogurt dip)

熱量 (Energy)	碳水化合物 (Carbohydrate)	蛋白質 (Protein)	脂肪 (Fat)	膳食纖維 (Dietary Fibre)
119	19	3	4	3
千卡 (kcal)	克 (g)	克 (g)	克 (g)	克 (g)

材料

- ☐ 馬鈴薯 200 克
- ☐ 蒜粉 1 茶匙（無鹽）
- ☐ 鹽 1/4 茶匙
- ☐ 紅椒粉 1 茶匙
- ☐ 巴馬臣芝士 1 湯匙
- ☐ 植物油 2 茶匙

低脂乳酪醬

- ☐ 低脂原味希臘乳酪 1/4 杯
- ☐ 檸檬汁 1-2 茶匙
- ☐ 蒜蓉 1 茶匙
- ☐ 細香葱 1 湯匙（切碎）
- ☐ 鹽 1/8 茶匙
- ☐ 黑椒碎 1/8 茶匙

做法

1. 馬鈴薯洗淨，切小塊成薯角，放入暖水浸 15 分鐘。預熱焗爐至 200℃。
2. 烤盤鋪上焗爐紙，備用。將低脂乳酪醬的材料放入大碗，拌勻，冷藏約 15 分鐘備用。
3. 將薯角表面的澱粉質沖走，印乾多餘水分。
4. 薯角放入食物盒，加入蒜粉、鹽、紅椒粉、巴馬臣芝士及植物油，不斷搖動至均勻。
5. 將薯角平放在預備好的烤盤，以 200℃焗 30-35 分鐘。想烤成均勻的效果，建議每 15 分鐘將薯角翻轉。

6. 焗好後，配上低脂乳酪醬即成。

小助手參與步驟

- 相比坊間高鹽分、多油的炸薯角，用少量油焗薯角的方式相對較健康，更能控制用油量及調味，食譜用了無鹽蒜粉代替部分鹽分，用油量相對較少，但始終是高溫處理，注意別將食物燒焦。
- 除了馬鈴薯，還可以焗其他根莖類食物，如番薯和紅蘿蔔。根莖類食物營養高，若連皮進食可攝取更多膳食纖維，也可間中代替麵包、白飯等。

|3 人分量|

野菜愛心玉子燒 Cheesy Rolled Omelette

每份 (per serving)

熱量 (Energy)	碳水化合物 (Carbohydrate)	蛋白質 (Protein)	脂肪 (Fat)	膳食纖維 (Dietary Fibre)
94	3	8	6	1
千卡 (kcal)	克 (g)	克 (g)	克 (g)	克 (g)

材料

- ☐ 雞蛋 3 隻
- ☐ 洋葱碎 2 湯匙
- ☐ 紅蘿蔔粒 3 湯匙
- ☐ 西蘭花粒 3 湯匙
- ☐ 低脂芝士 1 片
- ☐ 鹽及黑椒碎各適量
- ☐ 植物油 1-2 茶匙

做法

1. 雞蛋及其他材料放入大碗內（芝士除外），拂勻。
2. 預熱平底鍋，均勻地塗上少許油。倒入 1/3 杯蛋液，轉動平底鍋讓蛋液均勻佈滿鍋面。
3. 用中小火將蛋液煎至半熟，均勻地加入芝士片，將蛋皮捲成蛋卷。第一層完成後，再於鍋面抹上少量油及倒入適量蛋液（蛋液必須接觸已捲好的蛋卷）。
4. 重複以上步驟，直至所有蛋液用完，上碟，待涼，定型後斜切成心形。

小助手參與步驟

營養小貼士

- 雞蛋是一種價錢廉宜、容易購買的優質蛋白質來源。除了焓、蒸、炒的方式，還可以加入多種食材製作菜式，用途非常廣泛；但留意雞蛋要徹底煮熟，直至蛋黃及蛋白變得堅實，以減低引發食源性疾病的機會。
- 除了以上配搭，大家可加入其他健康食材，如原味紫菜、水浸吞拿魚、粟米粒等，製作多種口味。

| 4 人分量 |

非油炸脆脆雞塊

Baked Crispy Chicken Nuggets

每份 (per serving)
* 以半杯粟米片及不包括橄欖油
(using 1/2 cup of corn flakes, excluding olive oil)

熱量 (Energy)	碳水化合物 (Carbohydrate)	蛋白質 (Protein)	脂肪 (Fat)	膳食纖維 (Dietary Fibre)
138	16	14	2	1
千卡 (kcal)	克 (g)	克 (g)	克 (g)	克 (g)

材料

- ☐ 雞胸肉 200 克（切成一口大小，約 20 克）
- ☐ 原味脫脂乳酪 2 湯匙
- ☐ 原味粟米片 1/2 杯 -1 杯（壓碎）
- ☐ 巴馬臣芝士碎 2 湯匙
- ☐ 蒜粉 1/2 茶匙（無鹽）
- ☐ 鹽 1/8 茶匙
- ☐ 黑椒碎適量
- ☐ 橄欖油適量（噴霧裝）

做法

1. 預熱焗爐至攝氏 200℃；烤盤鋪上焗爐紙，備用。
2. 碗內加入雞肉、鹽、黑椒碎和乳酪拌勻，放於雪櫃醃 15 分鐘。
3. 於另一碗內，加入粟米片、巴馬臣芝士碎、蒜粉、鹽和黑椒碎拌勻。
4. 雞肉沾上粟米片和調味料，噴上適量橄欖油（可省略）。
5. 放入焗爐焗約 15-18 分鐘，至雞肉熟透及外層呈金黃，即成（想達到更均勻的效果，建議焗約 7 分鐘時翻轉雞塊再焗）。

營養小貼士

- 市面出售的脆雞塊很受歡迎，但一般用油炸方式製成，雞塊成分不明，脂肪含量也特別高。此食譜選用較瘦的雞胸肉為主要食材，並以烤焗的方式烹調，大大減低脂肪含量。
- 原味乳酪可配上水果、果仁直接食用，更可用來做成低脂醬料醃肉，作為天然的鬆肉材料。

| 4 人分量 |

日式拉絲三文魚金磚

Japanese Cheesy Salmon Cubes

每份 (per serving)

熱量 (Energy)	碳水化合物 (Carbohydrate)	蛋白質 (Protein)	脂肪 (Fat)	膳食纖維 (Dietary Fibre)
135	8	14	5	2
千卡 (kcal)	克 (g)	克 (g)	克 (g)	克 (g)

材料

- ☐ 馬鈴薯 1 個（中型）
- ☐ 急凍三文魚柳 150 克
- ☐ 雞蛋 1 隻
- ☐ 高鈣低脂芝士 2 片
- ☐ 鹽 1/4 茶匙
- ☐ 刁草 2 茶匙
- ☐ 麵粉 10 克
- ☐ 橄欖油 2 茶匙

做法

1. 馬鈴薯洗淨、去皮，切粒，放進沸水煮軟，盛起，瀝乾水分備用。
2. 三文魚以鹽和刁草醃約 15 分鐘，備用。
3. 易潔鑊內加入橄欖油 1 茶匙，以中火煎熟三文魚，用叉壓碎三文魚肉。
4. 馬鈴薯放碗內壓成蓉，加入三文魚碎和雞蛋混合。
5. 將三文魚薯蓉分成 16 等份，每份夾入少許高鈣低脂芝士並捏成長方狀，並灑上麵粉。
6. 將餘下的橄欖油加入易潔鑊，以中火將三文魚磚煎至每邊金黃色即成。

小助手參與步驟

營養小貼士

- 三文魚和雞蛋能提供蛋白質，有助修補身體組織和細胞，並分解成氨基酸製造白血球及抗體，以維持身體的免疫力。
- 三文魚含豐富健康多元不飽和脂肪——奧米加 3 脂肪酸，因未能透過身體大量製造，需從食物中攝取。多元不飽和脂肪有助降低體內的壞膽固醇水平，維持心血管健康。
- 美國心臟協會建議每星期進食兩次深海魚類（如：三文魚、吞拿魚等）以攝取奧米加 3 脂肪酸，更有研究顯示含奧米加 3 脂肪酸的深海魚類，對緩解抑鬱情緒有幫助。
- 不愛吃魚或素食者，可考慮進食奇亞籽、亞麻籽和合桃攝取奧米加 3 脂肪酸。

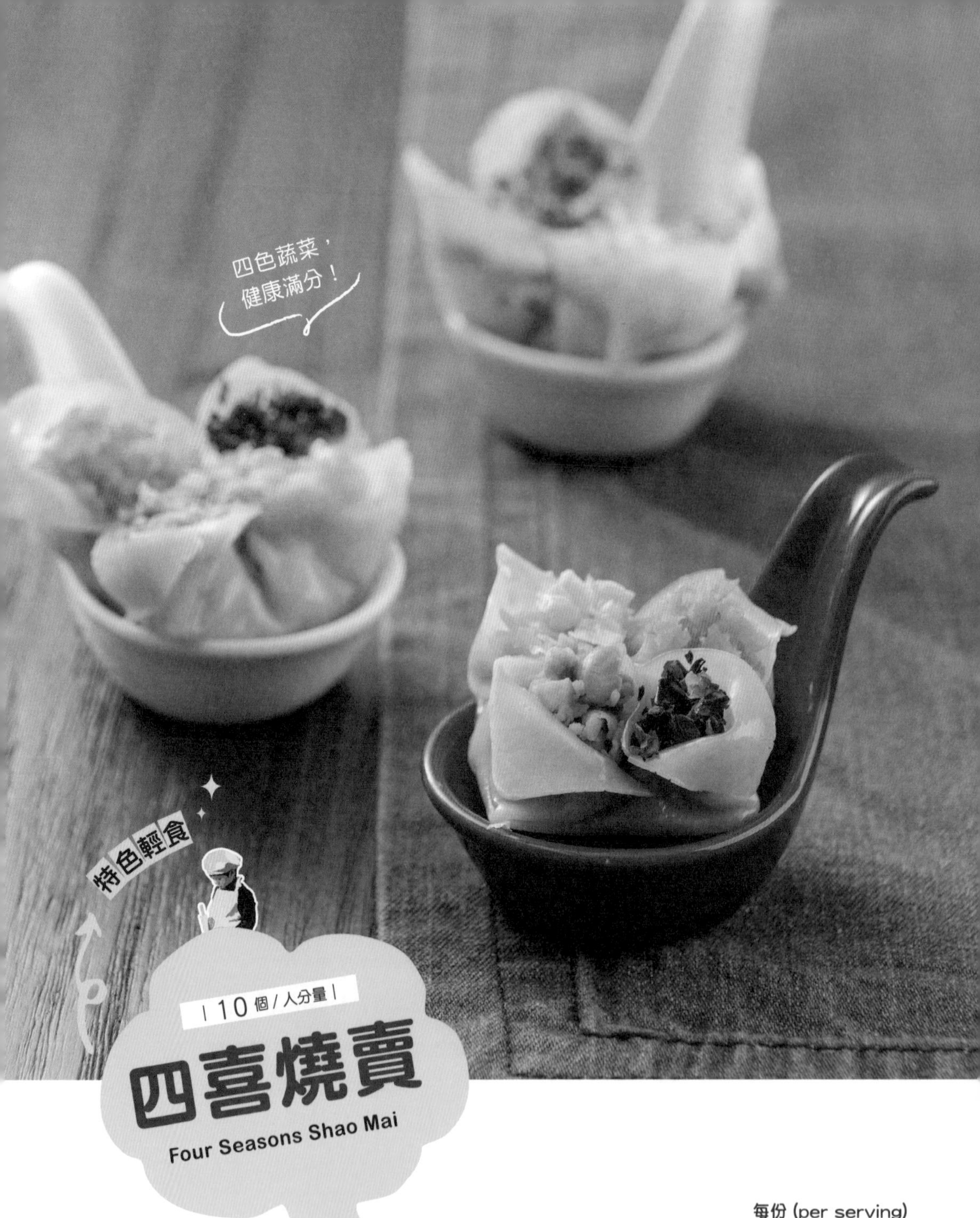

| 10 個 / 人分量 |

四喜燒賣
Four Seasons Shao Mai

每份 (per serving)

熱量 (Energy)	碳水化合物 (Carbohydrate)	蛋白質 (Protein)	脂肪 (Fat)	膳食纖維 (Dietary Fibre)
50	8	4	1	2
千卡 (kcal)	克 (g)	克 (g)	克 (g)	克 (g)

材料

- ☐ 方形餃子皮 10 片
- ☐ 免治瘦豬肉 100 克
- ☐ 雲耳 2 湯匙
- ☐ 甘筍 2 湯匙
- ☐ 粟米 2 湯匙
- ☐ 青豆 2 湯匙
- ☐ 生抽 2 茶匙
- ☐ 麻油 1/3 茶匙
- ☐ 薑蓉 1/4 茶匙

做法

1. 免治瘦豬肉加入薑蓉、生抽和麻油醃約 15 分鐘。
2. 雲耳、甘筍、粟米、青豆分別以沸水煮熟，放涼、切碎備用。
3. 餃子皮中央放入免治豬肉 1 茶匙，將餃子皮對摺，以少許水將長邊沿的中間點黏緊。
4. 將短邊的中間點摺向中央及按實，慢慢地打開四個缺口。
5. 在每個缺口分別放入各 1/2 茶匙雲耳、甘筍、粟米及青豆。
6. 燒沸水，排入四喜燒賣隔水蒸約 10 分鐘至熟透即成。

小助手參與步驟

四喜燒賣造型

營養小貼士

- 這個食譜利用不同顏色的蔬菜製作，吸引兒童進食的興趣。
- 四喜燒賣相比一般魚肉燒賣更健康，除利用新鮮豬肉製作，更加入不同的蔬菜增加膳食纖維攝取量。
- 蒸的烹調方法減少額外的脂肪攝取，屬健康的烹調方法。
- 按豬肉餡料的分量可調節燒賣大小，成為一道主菜或下午茶小食。

有營主菜・湯

熱量 (Energy)	碳水化合物 (Carbohydrate)	蛋白質 (Protein)	脂肪 (Fat)	膳食纖維 (Dietary Fibre)
287	38	15	10	5
千卡 (kcal)	克 (g)	克 (g)	克 (g)	克 (g)

每份 (per serving)

| 2 人分量 |

動物田園菠菜湯
Animal Garden Spinach Soup

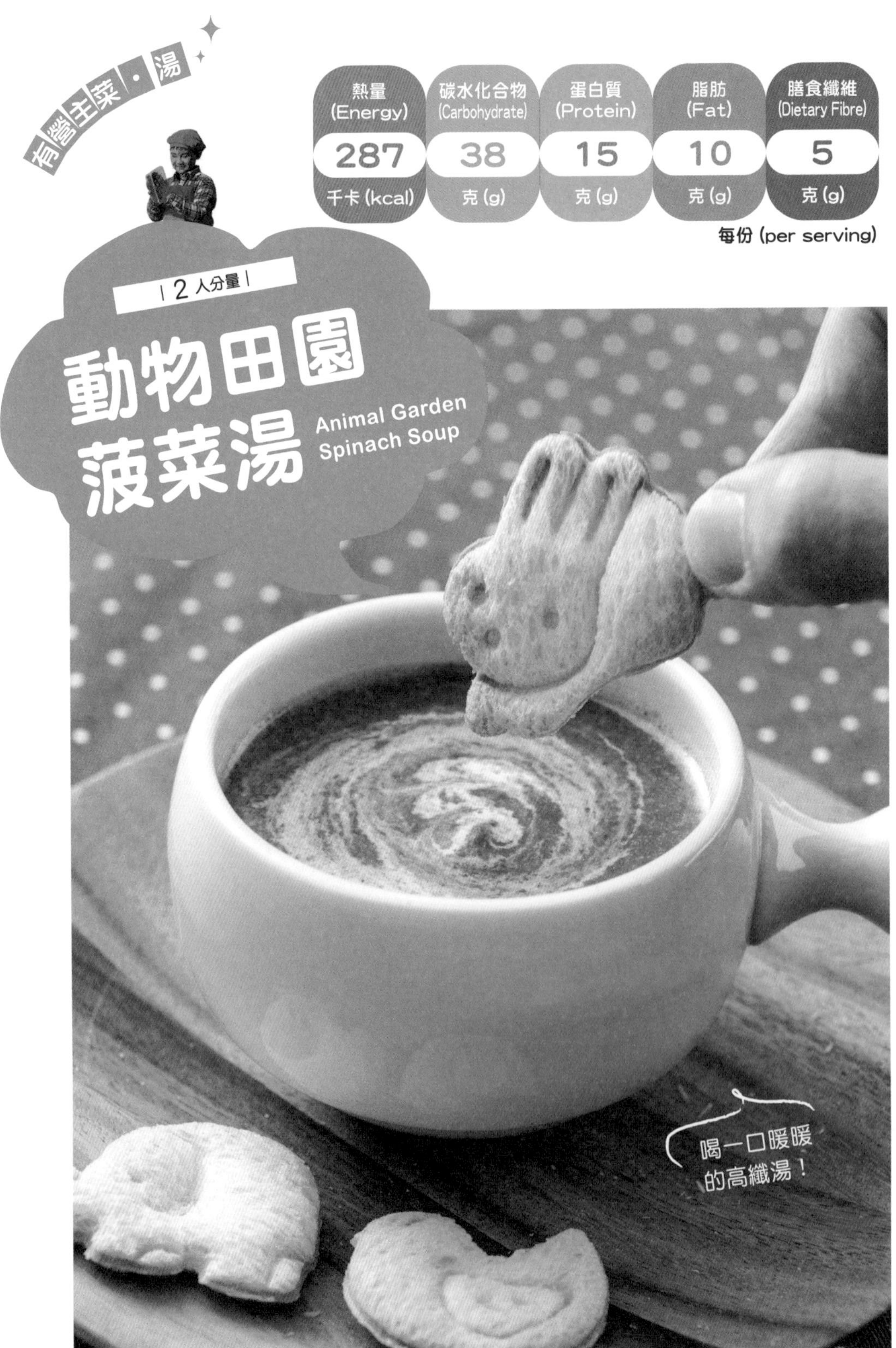

材料

- ☐ 方包 2 片
- ☐ 菠菜 250 克
- ☐ 馬鈴薯 1/2 個（中型）
- ☐ 洋葱 1/2 個（切碎）
- ☐ 蒜頭 1 瓣（切碎）
- ☐ 低脂鮮奶 300 毫升
- ☐ 低鈉雞湯 250 毫升
- ☐ 鮮檸檬汁 1/4 個
- ☐ 水適量
- ☐ 橄欖油 1 湯匙

裝飾

- ☐ 低脂鮮奶 15 毫升

做法

1. 用曲奇餅模具將方包印成不同的動物形狀，烘焗備用。
2. 馬鈴薯洗淨、去皮，切小塊備用。
3. 燒熱易潔鑊，加入油，下洋葱和蒜蓉以小火炒香，加入馬鈴薯塊略煎香，加入雞湯煮 10 分鐘至馬鈴薯軟身。
4. 加入低脂鮮奶、檸檬汁和半份菠菜，煮至軟身（約 15 分鐘），放涼約 5 分鐘。
5. 將湯倒進攪拌機，並加入餘下的菠菜攪拌，打至質地軟滑。
6. 湯加入鍋中煮熱（如湯太濃稠，可逐少加入水分調整至喜愛的濃稠度）。
7. 湯倒入碗內，最後加上麵包片及低脂鮮奶於湯面裝飾即成。

小助手參與步驟

營養小貼士

- 有時候，兒童的胃口不佳，食慾會減少，此食譜能幫助兒童攝取熱量和營養。
- 西式湯較中式湯水營養豐富，由於所有材料經攪拌後與湯水混合，因此每喝一口湯能攝取當中的營養素。
- 菠菜含有較豐富的鐵質和葉酸，每 100 克分別含有約 2.7 毫克鐵質和 194 微克葉酸，有助製造細胞和紅血球。
- 此食譜利用馬鈴薯製作，代替一般濃湯使用的忌廉，脂肪量減少之餘也同樣做到香濃軟滑的口感。

隨心滋味台式蛋餅

Build-your-own Taiwanese Egg Crepe

每份 (per serving)
* 不包括餡料 (excluding the fillings)

熱量 (Energy)	碳水化合物 (Carbohydrate)	蛋白質 (Protein)	脂肪 (Fat)	膳食纖維 (Dietary Fibre)
212	26	9	8	1
千卡 (kcal)	克 (g)	克 (g)	克 (g)	克 (g)

材料

- ☐ 雞蛋 3 隻

麵糊料

- ☐ 中筋麵粉 40 克
- ☐ 燕麥粉 40 克
- ☐ 粟粉 20 克
- ☐ 開水 200 克
- ☐ 葱適量（切細粒）
- ☐ 鹽、胡椒粉各適量
- ☐ 油 1-2 茶匙

醬汁

- ☐ 較低鈉醬油 1 湯匙
- ☐ 蒜蓉 1 茶匙
- ☐ 麻油 1 茶匙
- ☐ 開水 1/2 湯匙

餡料

- ☐ 水浸吞拿魚、低脂芝士、粟米粒、熟雞肉絲、熟蘑菇片（隨意配搭）

做法

❶ 麵糊料放入大碗拌勻，備用。另一個碗內，放入醬汁材料，拌勻備用。

❷ 平底鍋平均塗上少許油，倒入 1/3 杯麵糊，轉動平底鍋，盡量讓麵糊均勻佈滿整個鍋面。

❸ 餅煎熟後，加入雞蛋並輕輕撥開蛋液，繼續煎至熟透，鋪上自選餡料，將蛋餅捲起及塗上醬汁即成。

小助手參與步驟

營養小貼士

- 用攪拌機將燕麥攪拌成幼粉狀代替一般白麵粉，可增加蛋餅的膳食纖維量。
- 預備餡料時，多預備一些較有營養的食材給兒童，盡量避免使用加工肉類或醃菜。選擇罐頭吞拿魚時，多選較低鈉的水浸吞拿魚。
- 製作時，可加入不同顏色的蔬菜，豐富菜式的色彩。
- 利用天然調味料如蒜頭，可提升菜式的味道，並減少現成醬汁的使用。當使用現成醬汁時，不妨閱讀營養標籤，選擇較低鈉的醬汁。

| 4 人分量 |

鮮蝦飯糰波波

Shrimp Rice Balls

每份 (per serving)

熱量 (Energy)	碳水化合物 (Carbohydrate)	蛋白質 (Protein)	脂肪 (Fat)	膳食纖維 (Dietary Fibre)
59	9	3	1	1
千卡 (kcal)	克 (g)	克 (g)	克 (g)	克 (g)

材料

- ☐ 蝦仁 4 隻
- ☐ 燕麥飯 2/3 碗
- ☐ 壽司醋 1 湯匙
- ☐ 粟米粒 1 湯匙
- ☐ 紅蘿蔔粒 1 湯匙
- ☐ 壽司紫菜碎 2 湯匙
- ☐ 植物油 1/2 茶匙
- ☐ 鹽 1/8 茶匙
- ☐ 胡椒粉適量

做法

❶ 蝦仁用鹽和胡椒粉醃 15 分鐘，蒸熟備用。

❷ 平底鍋加入 1/2 茶匙植物油，將紅蘿蔔粒炒熟，取出。

❸ 大碗內放入燕麥飯、壽司醋、粟米粒、紅蘿蔔粒及壽司紫菜碎，拌勻。

❹ 預備一張保鮮紙，先放上蝦仁（讓蝦仁置於飯糰表面），再鋪上燕麥飯，搓成球狀即成。

小助手參與步驟

營養小貼士

- 飯糰加入了燕麥米和多種蔬菜，能增添口感和色彩之外，還有助提高膳食纖維含量。
- 蝦仁是提供蛋白質的食物，其脂肪含量較低，4 隻中型蝦仁相等一份肉類。
- 一般日式壽司選用刺身作食材，未經煮熟的肉類較有機會帶有食源性致病菌，或令兒童有致病的風險，因此不適合兒童食用。
- 可按兒童的喜好，自行挑選飯糰配料，如雞肉或三文魚塊代替蝦仁。

| 5 人分量 |

豆腐煎藕餅

Pan-fried Tofu Lotus Root Patty

每份 (per serving)

熱量 (Energy)	碳水化合物 (Carbohydrate)	蛋白質 (Protein)	脂肪 (Fat)	膳食纖維 (Dietary Fibre)
146	8	12	8	2
千卡 (kcal)	克 (g)	克 (g)	克 (g)	克 (g)

材料

- ☐ 硬豆腐 200 克
- ☐ 免治雞肉 150 克
- ☐ 雞蛋 1 隻
- ☐ 甘筍 2 湯匙（切粒）
- ☐ 洋葱 2 湯匙（切粒）
- ☐ 麵包糠 20 克
- ☐ 蓮藕 10 片
- ☐ 芥花籽油 2 茶匙

裝飾

- ☐ 毛豆適量

調味料

- ☐ 鹽 1/8 茶匙
- ☐ 糖 1/2 茶匙
- ☐ 生抽 1 茶匙
- ☐ 麻油 1 茶匙

做法

1. 硬豆腐以廚房紙輕輕按壓，吸走水分。
2. 硬豆腐壓碎，逐一加入免治雞肉、雞蛋、甘筍粒、洋葱粒、麵包糠和調味料，混合備用。
3. 易潔鑊加入油，燒熱後將混合的豆腐雞肉分成 10 份，放進鑊內按壓成圓形。
4. 以中火慢煎至半熟，並將蓮藕片置於每個肉餅上，加上毛豆裝飾，翻轉另一邊，煎至肉餅熟透即成。

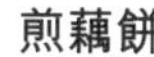

- 豆腐和雞肉質地軟身，兒童較易進食。雖然蓮藕片質地較硬，也可訓練兒童的咀嚼能力，家長可按照兒童日常的咀嚼程度，調節蓮藕片的厚薄。
- 蓮藕提供膳食纖維，每 100 克有約 5 克膳食纖維，一般用蓮藕製作的傳統菜式未必吸引兒童嘗試，不妨利用此食譜讓兒童接觸新食材。
- 食用時可添加沙律菜裝飾伴食，增加口感和清新感覺，更可藉此增加膳食纖維的攝取，有助腸臟蠕動，預防便秘。
- 兒童幫忙按壓豆腐，並混合食物材料，令他們對食物產生興趣，亦可藉此作為兒童的手部活動訓練。

日式茶碗蒸

Japanese Steamed Egg

每份 (per serving)

熱量 (Energy)	碳水化合物 (Carbohydrate)	蛋白質 (Protein)	脂肪 (Fat)	膳食纖維 (Dietary Fibre)
88	3	11	4	1
千卡 (kcal)	克 (g)	克 (g)	克 (g)	克 (g)

材料

- ☐ 急凍雞腿肉 120 克
- ☐ 急凍蝦仁 4 隻
- ☐ 鮮冬菇 4 朵（小型）
- ☐ 秋葵 2 條
- ☐ 甘筍 8 片
- ☐ 雞蛋 2 隻
- ☐ 生抽 1/4 茶匙
- ☐ 熱水 300 毫升
- ☐ 鰹魚粉 1 茶匙

調味料

- ☐ 鹽 1/4 茶匙
- ☐ 糖 1/2 茶匙
- ☐ 生抽 1 茶匙

做法

❶ 雞腿肉洗淨、去皮和脂肪，以廚房紙抹乾，切小塊，加入生抽醃約 15 分鐘，備用。

❷ 蝦仁洗淨，以廚房紙抹乾備用。

❸ 冬菇抹乾，在菇面切成花形圖案；秋葵切成星狀；甘筍片以模具印出圖案，與秋葵以沸水煮軟，放涼備用。

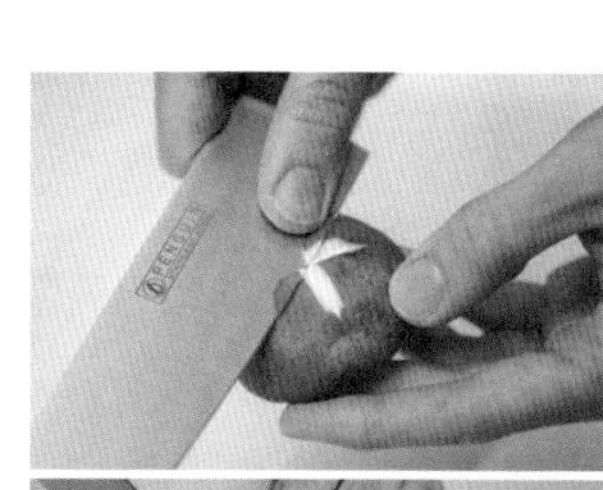

❹ 鰹魚粉混合熱水和調味料，待冷後加入雞蛋拂打。

❺ 茶杯中放入雞肉和蝦仁，以濾網慢慢地倒入鰹魚蛋汁，放入冬菇、甘筍和秋葵於蛋面。

❻ 以保鮮紙蓋好（不用密封），以防蒸煮時水分滴到蒸蛋破壞美感，以小火隔水蒸約 10 分鐘即成。

小助手參與步驟

- 急凍雞肉和蝦仁可預先購買並存放冰箱，方便隨時加入菜式，兩者能提供優質蛋白質；去皮的雞肉與蝦仁屬低脂食材。
- 茶碗蒸加入不同種類、形狀和顏色的蔬菜，有助吸引兒童增加蔬菜的攝取，有助攝取膳食纖維，預防便秘。
- 利用鰹魚粉加入菜式，能提升菜式的鮮味，並減少糖和鹽的使用量。茶碗蒸以蒸的健康烹調方法製作，減少油分使用。
- 除了當正餐菜式，亦可當作小食以提供兒童蛋白質攝取。部分兒童因肉類質地太硬而產生抗拒，此菜式的雞肉和蝦質地軟身，配上香滑的蒸蛋能吸引兒童食用。

營養師聊天室

選哪款肉類最好？

蛋白質有助細胞痊癒及修復，並於體內分解成氨基酸，其中一些必需的氨基酸有助增加細胞免疫力機能。

海產類食物、低脂肉類、去皮家禽、雞蛋、乾豆、非油炸豆類製品及非油炸原味果仁能提供豐富的蛋白質，家長不妨考慮提供多種蛋白質來源，相對來說紅肉能提供較多鐵質；白肉則含較低脂肪；海產類食物含較豐富鋅質；深海魚類多含奧米加 3 脂肪酸；乾豆和果仁含較豐富膳食纖維；非油炸豆類製品質地較軟身，容易食用。

因此，沒有一款肉類代替品屬最好選擇，家長因應食材配搭、個人口味和營養素所需決定一週的肉類分配，以培養兒童多元化的飲食習慣，並達至均衡飲食的大原則。

| 3 人分量 |

鮮味大阪燒

Vegetable Okonomiyaki

每份 (per serving)
* 不包括海鮮配料 (excluding seafood)

熱量 (Energy)	碳水化合物 (Carbohydrate)	蛋白質 (Protein)	脂肪 (Fat)	膳食纖維 (Dietary Fibre)
133	18	4	4	2
千卡 (kcal)	克 (g)	克 (g)	克 (g)	克 (g)

材料

- ☐ 低筋麵粉 50 克
- ☐ 鰹魚碎 0.5 克
- ☐ 山藥 25 克（磨蓉）
- ☐ 椰菜絲 100 克
- ☐ 雞蛋 1 隻（中型）
- ☐ 鹽適量
- ☐ 植物油 1 茶匙

海鮮配料

- ☐ 帶子、蝦仁、魷魚（隨意）

調味料

- ☐ 低脂沙律醬 1 湯匙
- ☐ 大阪燒醬 1 湯匙
- ☐ 紫菜碎適量
- ☐ 鰹魚碎適量

做法

1. 大碗內加入低筋麵粉、鹽、鰹魚碎拌勻。
2. 山藥泥、雞蛋、海鮮和椰菜絲拌勻備用。
3. 燒熱平底易潔鑊，加入油，倒入半份麵糊，用中火煎 5 分鐘（期間翻轉）至全熟及呈金黃色，抹上沙律醬及大阪燒醬，最後撒上紫菜碎及鰹魚碎即成。

小助手參與步驟

營養小貼士

- 一般大阪燒會加入較高脂肪的豬肉片，這個食譜用了較低脂的海鮮作配料，令菜式更健康之餘，也不失鮮味！
- 用易潔鑊烹調，有助控制用油量。

兒童有食物過敏問題，如何處理？

食物過敏可大可小，當有食物過敏的兒童接觸致敏源，身體或會出現各種過敏反應，如皮膚出疹、紅腫、嘔吐、腸胃不適、心跳加速，甚至出現呼吸困難或休克致命。較常見的致敏源包括雞蛋、海產、牛奶、花生、果仁、大豆、小麥、一些食物添加劑等。

如有任何懷疑個案，家長應及早就醫，請醫生透過過敏源測試診斷，一經確診，應盡快通知學校作出特別膳食安排。一般可以透過避免進食含有致敏源的食物，減少及避免過敏反應帶來的影響。日常飲食時建議如下：

- 家長及兒童多留意包裝食物上的成分表及認清致敏物來源，尤其隱藏於加工食物的致敏物。
- 不應胡亂及過度戒口，以免影響營養吸收及生活質素。
- 記得以適當的食物補充因戒口而受影響的營養素。
- 留意並減低製作及包裝食物過程中，受致敏源交叉污染的機會。
- 定期覆診，多監察兒童的營養攝取及成長狀況，如有任何疑問諮詢註冊營養師的意見。

另外，家長亦常問到：「如孩子有乳糖不耐症，應否避免飲奶？」

其實乳糖不耐症是指體內缺乏乳糖酵素，以致難以分解奶製品中的乳糖，從而引致腹痛、腹瀉等不適徵狀。症狀由輕微至嚴重，視乎患者接受乳糖的程度。

乳糖不耐症並非食物過敏的一種，所以跟食物敏感的處理方法有別。家長可以嘗試將少量含有乳糖的食物與其他食物讓兒童一起進食，令身體學習及適應，或許可以克服乳糖不耐症。

| 2 人分量 |

夏日彩蔬冷麵

Rainbow Veggie Noodle Salad

每份 (per serving)

熱量 (Energy)	碳水化合物 (Carbohydrate)	蛋白質 (Protein)	脂肪 (Fat)	膳食纖維 (Dietary Fibre)
284	42	16	8	2
千卡 (kcal)	克 (g)	克 (g)	克 (g)	克 (g)

材料

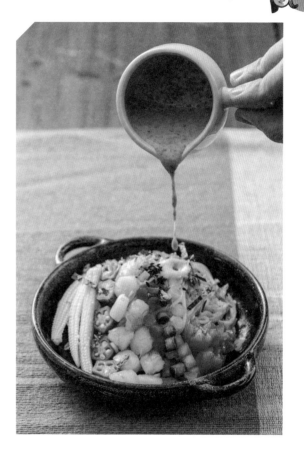

- ☐ 素麵 80 克
- ☐ 三色甜椒各 1/4 個
- ☐ 秋葵 2 條
- ☐ 車厘茄 4 顆
- ☐ 粟米仔 4 條
- ☐ 急凍蝦 4 隻
- ☐ 急凍帶子 4 隻
- ☐ 低脂芝麻醬 4 茶匙
- ☐ 黑芝麻 1 茶匙
- ☐ 原味紫菜碎適量
- ☐ 木魚碎適量

做法

❶ 素麵放於沸水煮熟（按包裝上的時間烹調），放涼備用。

❷ 三色甜椒切粒，或以模具切成不同形狀；秋葵切成星型；粟米仔切條。將以上食材汆水備用；車厘茄切半備用。

❸ 蝦及帶子解凍，汆水後切成合適的粒狀，備用。

❹ 麵條放於碟上，加上已煮熟的食材，最後淋上低脂芝麻醬，灑上黑芝麻、適量原味紫菜碎和木魚碎即成。

小助手參與步驟

- 炎熱的夏天有機會令兒童食慾大減，製作一個透心涼的菜式，不但幫助身體降溫，更能增加食慾，幫助攝取足夠的營養。
- 急凍蝦及帶子可常備於冰箱隨時使用，兩者為低脂食材及提供蛋白質。約 4-5 隻中型蝦或帶子計算，相等於 1 份肉類。
- 選用不同顏色及形狀的蔬菜，能吸引兒童進食，提供不同的植物元素，達到抗氧化作用。
- 購買低脂芝麻醬時應閱讀營養標籤，或可使用醋、新鮮檸檬汁代替，進一步減低脂肪和鈉質攝取。

營養師聊天室

有機食物是最好的選擇嗎？

家長當然將最好的留給子女，無論是學業還是衣食住行各方面，都希望盡量滿足子女的需要。現今社會，吃得足夠似乎已經未能達至家長心目中的「最好」，所以不時聽到家長問及「應否選擇有機食物？」

「有機」一詞風靡一時，無論附加於任何產品身上，一聽到「有機」二字會令大家覺得是天然、較好的選擇。在保護環境方面，有機種植相比農藥種植的確更能保護環境；但在營養角度方面，其實有機與否其營養價值相若，亦暫無完整的大型研究顯示利用有機種植的食物含有較豐富的營養素。

有機食材除價格較昂貴，經烹調後同樣可以變為不健康的食物，例如：天婦羅有機秋葵（大大增加油分攝取）。因此，家長必須小心留意食物的選擇，按自己的金錢預算決定是否購買有機食材，外出進食時亦別輕易被「有機」一詞健康化該食物。

|4人分量|

小鹿造型醬汁牛肉卷餅 Reindeer Beef Rolls

每份 (per serving)

熱量 (Energy)	碳水化合物 (Carbohydrate)	蛋白質 (Protein)	脂肪 (Fat)	膳食纖維 (Dietary Fibre)
199	26	9	8	3
千卡 (kcal)	克 (g)	克 (g)	克 (g)	克 (g)

材料

- ☐ 墨西哥薄餅 2 片
- ☐ Pretzel 脆餅 8 塊
- ☐ 牛柳 80 克
- ☐ 牛油生菜 1/4 個（細）
- ☐ 甘筍 1/2 條
- ☐ 紫洋葱 1/2 個
- ☐ 車厘茄 4 顆
- ☐ 蒜蓉 1/4 茶匙
- ☐ Mozzarella 芝士碎 20 克
- ☐ 牛油果 1/2 個
- ☐ 新鮮檸檬汁 1/2 茶匙
- ☐ 芥花籽油 1 茶匙

調味料

- ☐ 鹽 1/8 茶匙
- ☐ 糖 1/2 茶匙
- ☐ 粟粉 1/2 茶匙
- ☐ 意式香草適量

做法

1. 牛柳洗淨，以廚房紙抹乾，切絲，加入調味料醃 15 分鐘備用。
2. 牛油生菜、甘筍、紫洋葱和車厘茄洗淨；牛油生菜撕成片；甘筍及紫洋葱切幼絲；車厘茄切半備用。
3. 易潔鑊下油燒熱，加入紫洋葱和牛柳絲以中火炒熟，期間加入蒜蓉炒香。
4. 牛油果加入檸檬汁壓蓉；Pretzel 脆餅切半。
5. 墨西哥薄餅鋪好，塗上牛油果醬，依次加上牛油生菜、甘筍絲、紫洋葱絲和牛柳絲，最後加入芝士碎。
6. 墨西哥薄餅捲好，切成 4 件，以車厘茄和 Pretzel 脆餅裝飾成小鹿造型即成。

小助手參與步驟

牛肉卷餅

營養小貼士

- 牛肉除了提供蛋白質，更含有豐富的鐵質，能製造血紅素形成紅血球，將氧氣輸送到身體各組織。如鐵質攝取不足會引致缺鐵性貧血。
- 缺鐵性貧血會令臉色蒼白、容易疲勞，學習時難以集中精神。此外，亦會引起頭暈、食慾不振、身體抵抗力下降等健康問題。
- 鐵質需求會因年齡和性別而有所不同，應向醫生或註冊營養師查詢。以4至6歲男女學童為例，每天需攝取10毫克鐵質。
- 除紅肉之外，雞蛋、乾果、深綠色蔬菜均提供鐵質。植物性食物提供的鐵質則需維他命C幫助吸收，如番茄、檸檬汁等。

營養師聊天室

零食？小食？你吃對了嗎？

零食與小食的分別在於其熱量和營養的供應和進食時間。

零食，多為高熱量、低營養價值的食物，如糖果、薯片等，通常沒有既定的食用時間，想吃就吃、好心情就吃，有些家長更以零食作獎勵。過量進食零食會影響食用正餐的胃口，更會引致肥胖問題。

相反，小食則提供熱量和營養素的健康食物，多為天然的食物，如新鮮水果、車厘茄、原味乳酪等。通常因感到肚餓才進食，並安排於兩餐正餐之間，如上午茶點、下午茶點。由於兒童的胃口較少且活動量較大，適當及適量地進食健康小食有助兒童攝取足夠的營養，或補償正餐時吃不足夠的熱量。

不過，家長要謹記需以不影響正餐進食的分量為大前提，並着重「質」和「量」，選擇低油、低鹽、低糖和高纖維的健康小食，例如新鮮水果、半杯低脂奶、半份雞蛋三文治等。當然，可參考本書不同的健康小食，讓兒童享受食物之時，也攝取足夠的熱量和營養素。

| 4 人分量 |

迷你快「酪」PIZZA

Greek Yogurt Pizza Crust

每份 (per serving)

*使用 2 湯匙番茄醬 (using 2 tbsp of tomato sauce)

熱量 (Energy)	碳水化合物 (Carbohydrate)	蛋白質 (Protein)	脂肪 (Fat)	膳食纖維 (Dietary Fibre)
171	23	12	3	1
千卡 (kcal)	克 (g)	克 (g)	克 (g)	克 (g)

餅底材料

- ☐ 自發粉 110 克
- ☐ 低脂希臘乳酪 1/2 杯

配料

- ☐ 低鈉番茄醬 2-4 湯匙
- ☐ 菠菜苗 1/4 杯
- ☐ 蘑菇片 1/4 杯
- ☐ 熟雞肉絲 1/4 杯
- ☐ 粟米 2 湯匙
- ☐ 車厘茄 1/4 杯（洗淨、切半）
- ☐ 低脂芝士 2 片

做法

1. 將烤箱預熱至 200℃。
2. 自發粉和希臘乳酪放於大碗混合，用力揉成麵糰至光滑。
3. 麵糰放於已撒麵粉的桌面，一分為二，用擀麵棒擀成約 9 厘米圓餅（若麵糰黏着，可撒些麵粉），小心地放到已撒粉的烤盤上。
4. 番茄醬均勻地抹上麵糰，均勻地擺放配料，讓兒童發揮創意，做出自己心目中的口味，最後撒上低脂芝士。
5. 放入焗爐烤 20 分鐘，見麵糰邊成金黃色及芝士融化即成，切片享用！

小助手參與步驟

營養小貼士

- 自家製的薄餅口味可以千變萬化，可選自己喜愛的材料，做出屬於自己的作品。兒童也可以為薄餅改名，增加他們的滿足感。
- 如沒有雞肉，不妨用新鮮蝦仁、魚類或魷魚，做出不同風味的薄餅來。
- 除了親自製作薄餅底，大家可選用現成的全麥彼得包、全麥英式鬆餅、墨西哥薄餅，甚至全麥多士作餅底，方便又健康！

熱量 (Energy)	碳水化合物 (Carbohydrate)	蛋白質 (Protein)	脂肪 (Fat)	膳食纖維 (Dietary Fibre)
146	19	10	4	2
千卡 (kcal)	克 (g)	克 (g)	克 (g)	克 (g)

每份 (per serving)

| 2 人分量 |

菠菜烤闊麵條

Spinach and Tomato Lasagna

材料

- ☐ 免煮千層麵皮 35 克
- ☐ 菠菜苗 80 克
- ☐ 低鈉番茄醬 1/4 杯
- ☐ 低脂茅屋芝士 3 湯匙
- ☐ 巴馬臣芝士碎 2 湯匙
- ☐ 雞蛋 1/2 隻（中型）
- ☐ 鹽適量
- ☐ 黑椒碎適量

做法

1. 預熱焗爐至 180℃。菠菜苗煮熟，瀝乾水分，切碎備用。
2. 菠菜碎、低脂茅屋芝士、雞蛋、鹽和黑椒碎放進碗內，拌勻備用。
3. 焗盤內先加入 2 湯匙番茄醬，鋪上一層千層麵皮，再鋪上菠菜芝士混合物，鋪好 3 層後，用剩餘的番茄醬鋪滿麵條，均勻地灑上巴馬臣芝士。
4. 用錫紙蓋好，放入焗爐焗 30-40 分鐘，即可享用。

小助手參與步驟

營養小貼士

- 急凍蔬菜較新鮮的更有營養，保留了較多維他命及礦物質，而且貯藏方便，食譜內可選用急凍菠菜碎代替新鮮菠菜。選購急凍食品前，要留意營養標籤及食物成分，避免選購添加了鹽、糖及油的產品。
- 低脂茅屋芝士是營養豐富的小吃，能提供蛋白質，日常烹調可代替高脂肪芝士等。

每份 (per serving)

熱量 (Energy)	碳水化合物 (Carbohydrate)	蛋白質 (Protein)	脂肪 (Fat)	膳食纖維 (Dietary Fibre)
168	30	4	4	3
千卡 (kcal)	克 (g)	克 (g)	克 (g)	克 (g)

薯仔意粉材料

- ☐ 馬鈴薯 450 克（蒸熟，壓成蓉）
- ☐ 中筋麵粉 1 杯
- ☐ 鹽 1/4 茶匙

簡易番茄醬

- ☐ 洋葱 1 個（中型，切粒）
- ☐ 蒜蓉 1 茶匙
- ☐ 橄欖油 1 湯匙
- ☐ 低鈉番茄蓉 1 罐（400 克）
- ☐ 意大利雜香草 1/2 -1 茶匙（不含鹽）
- ☐ 鹽適量

配料

- ☐ 洋葱 1/4 個（中型，切粒）
- ☐ 車厘茄 1/4 杯（洗淨，切半）
- ☐ 巴馬臣芝士 3 湯匙
- ☐ 橄欖油 1 湯匙
- ☐ 羅勒葉適量

做法

1. 製作簡易番茄醬：加熱平底鍋，加入橄欖油，以慢火煮熟洋葱，加入其餘材料，用慢火煮至醬汁濃稠（約 20 分鐘），期間不停攪拌，備用。
2. 薯蓉、中筋麵粉及鹽放入碗內，輕輕搓成麵糰。
3. 在工作枱灑上少量麵粉，將麵糰分成 8 小份，慢慢搓成長幼條狀，切成小塊（成人拇指大小），用叉子壓出花紋，捲起備用。
4. 用中火煮滾一鍋水，加入薯蓉麵糰至浮起，慢慢撈起瀝乾，備用。
5. 加熱平底鍋，加少許橄欖油，爆香洋葱，加入已瀝乾薯蓉丸子煎至金黃色，下車厘茄快炒，加入約 1-2 杯番茄醬拌勻，下羅勒葉和巴馬臣芝士即成。

小助手參與步驟

營養小貼士

- 番茄味道濃郁，營養豐富，是維他命 A、C 和茄紅素的食物來源之一，用少許油加熱番茄能幫助茄紅素吸收。
- 新鮮和乾香草為天然調味料，適合用於多種菜式。以羅勒葉為例，特別適合用於番茄和意粉菜式上。
- 想薯仔丸子顏色更豐富，不妨加入適量的菠菜汁或紅菜頭汁等天然食用色素。

| 8 個 / 人分量 |

花花餃子

Rose Dumplings

每份 (per serving)

熱量 (Energy)	碳水化合物 (Carbohydrate)	蛋白質 (Protein)	脂肪 (Fat)	膳食纖維 (Dietary Fibre)
124	16	11	2	1
千卡 (kcal)	克 (g)	克 (g)	克 (g)	克 (g)

材料

- ☐ 免治瘦豬肉 300 克
- ☐ 即食燕麥片 1 湯匙
- ☐ 急凍三色雜豆 2 湯匙
- ☐ 圓形餃子皮 24 塊
- ☐ 水適量

調味料

- ☐ 麻油 1 茶匙
- ☐ 鹽 1/4 茶匙
- ☐ 糖 1/4 茶匙
- ☐ 生抽 1/2 茶匙

做法

1. 免治瘦豬肉以調味料醃 15 分鐘。
2. 三色雜豆解凍後，切碎。
3. 已醃好的豬肉、三色雜豆和即食燕麥片混合，以保鮮紙包好，放雪櫃冷藏 15 分鐘。
4. 將 3 片餃子皮橫向排列，並將 1/3 部分重疊，以水黏好。
5. 將約 1 湯匙餡料置於餃子皮中間，餃子皮從下而上以清水摺合成半圓形。餃子從左至右捲成花形，以水黏緊末端。
6. 鍋內燒熱滾水，隔水蒸餃子 15 分鐘即成。

小助手參與步驟

花花餃子造型

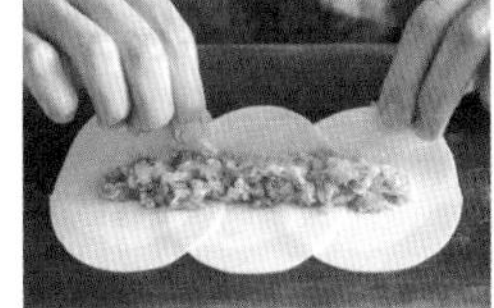
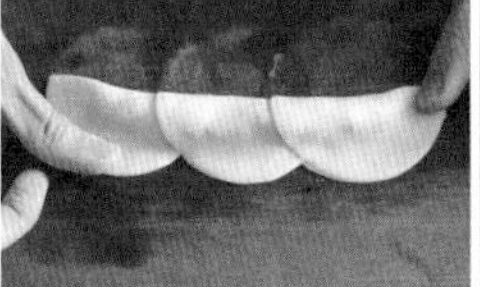
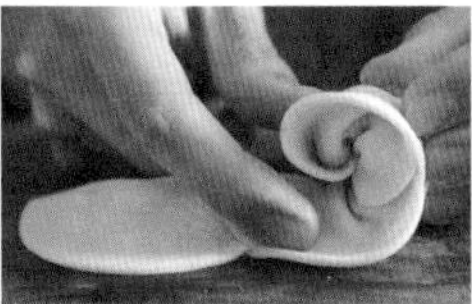

營養小貼士

- 雖然傳統餃子使用大量蔬菜製作，但大部分蔬菜需預先煮熟、隔水、放涼才混合肉類製成餡料，因此較費時。此食譜製作簡單，三色雜豆更可存放冰箱方便隨時使用。
- 即食燕麥片能吸收餃子餡料多餘的水分，更提供水溶性纖維素，有助心血管健康。
- 購買豬肉時，謹記購買瘦肉以減少飽和脂肪的攝取。餡料加入即食燕麥片，有助餡料質地變軟。
- 包好的餃子可存放冰箱數天，以蒸或煮的方式配以麵條和蔬菜，製成方便簡易的一餐。

有營主菜・湯

熱量 (Energy)	碳水化合物 (Carbohydrate)	蛋白質 (Protein)	脂肪 (Fat)	膳食纖維 (Dietary Fibre)
70	2	10	2	1
千卡 (kcal)	克 (g)	克 (g)	克 (g)	克 (g)

每份 (per serving)

| 5 人分量 |

小小雞肉紫菜卷

Seaweed Chicken Rolls

香脆有營的午餐之選！

材料

- ☐ 原味紫菜 2 大片（每片切成 10 等份）
- ☐ 植物油 1 茶匙

餡料

- ☐ 雞胸肉 200 克（攪碎）
- ☐ 椰菜 180 克（切碎、瀝乾）
- ☐ 洋葱 20 克（切碎）
- ☐ 鹽 1/2 茶匙
- ☐ 胡椒粉適量
- ☐ 麻油 1 茶匙

做法

1. 將所有餡料放入大碗內，用筷子攪拌至完全均勻及起膠。
2. 在每一塊小紫菜鋪上約 20 克餡料，捲起，做成一小卷紫菜雞肉卷。
3. 燒熱鑊，加入植物油，用慢火煎紫菜雞肉卷至金黃熟透，即可享用。

小助手參與步驟

營養小貼士

- 雞胸肉脂肪含量較低，攪碎後加入適量的蔬菜可讓口感、味道和營養更豐富。
- 大家可因應自己的口味，加入其他蔬菜如紅蘿蔔、白蘿蔔、香菇、香葱、粟米粒，讓兒童嘗試多種食材，攝取更多元化的營養。

每份 (per serving)

熱量 (Energy)	碳水化合物 (Carbohydrate)	蛋白質 (Protein)	脂肪 (Fat)	膳食纖維 (Dietary Fibre)
211	25	13	7	3
千卡 (kcal)	克 (g)	克 (g)	克 (g)	克 (g)

為公主設計衣裳吧！

有營主菜・湯

| 4 人分量 |

公主裙蛋包飯

Princess Omurice Gown

材料

- ☐ 免治牛肉 100 克
- ☐ 洋葱 1/4 個
- ☐ 蒜蓉 1 茶匙
- ☐ 蘑菇 5 顆
- ☐ 罐頭番茄蓉 200 克
- ☐ 急凍三色雜豆 50 克
- ☐ 白飯或燕麥飯 1 碗
- ☐ 雞蛋 2 隻
- ☐ 低脂鮮奶 1 湯匙
- ☐ 芥花籽油 3 茶匙
- ☐ 糖 1/4 茶匙
- ☐ 鹽 1/4 茶匙

調味料

- ☐ 生抽 1 茶匙
- ☐ 粟粉 1 茶匙
- ☐ 糖 1/2 茶匙

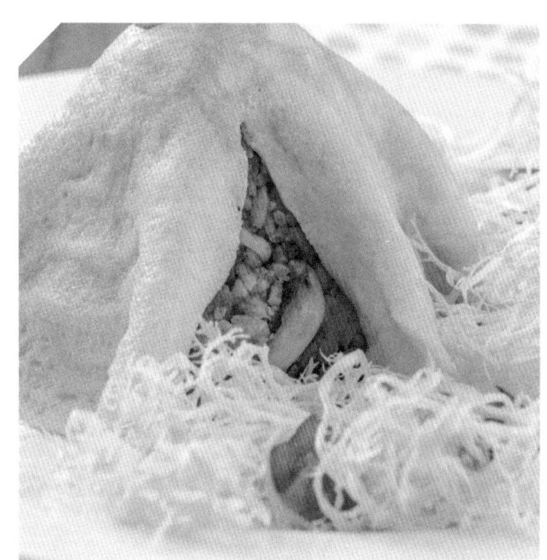

做法

1. 免治牛肉加入調味料，醃約 15 分鐘備用。
2. 洋葱切碎；蘑菇切片；三色雜豆略洗，去掉冷藏味，瀝乾備用。
3. 燒熱鑊，加入油 2 茶匙，下洋葱炒至軟身，加入蒜蓉炒香。下蘑菇片和免治牛肉炒至八成熟，加入番茄蓉以慢火煮至醬汁濃稠。
4. 加入三色雜豆、白飯、糖和鹽炒勻，放入兩個碗內並倒轉置於碟上。
5. 以保鮮紙按成公主裙形狀。
6. 雞蛋與低脂鮮奶拌勻，以濾網隔走蛋筋。
7. 以中火燒熱鑊，加入油 1 茶匙，倒進蛋液煎成圓形，加蓋後關火焗熟。
8. 蛋皮放在炒飯上，以蔬菜裝飾裙腳，於蛋包飯頂部加上公主卡紙即成。

小助手參與步驟

- 雞蛋與牛肉屬肉類食物，為身體提供蛋白質。一隻雞蛋的營養與一份肉類相等，質地軟身較易讓兒童咀嚼，確保攝取足夠的蛋白質，有助肌肉成長。
- 選擇罐頭番茄蓉應留意包裝上的食材表和營養標籤，有助選擇較低糖分和鈉質的產品。如選擇已添加糖或鹽的罐頭番茄蓉，炒飯時的糖和鹽可省略，以控制兩者的攝取量。
- 食譜內的免治牛肉可換成去皮雞柳、紅腰豆等提供蛋白質的食材。裝飾用的蔬菜可按兒童的喜好轉換，如甘筍和西蘭花，增添食用興趣。

得意食譜改善偏食和挑食習慣

兒童出現過瘦或營養不良，有機會來自偏食或挑食的飲食習慣。

偏食，即指完全不進食某一大類別的食物，導致未能攝取該食物組別的營養素，例如完全不進食任何水果，導致膳食纖維和水溶性維他命缺乏。

挑食，即揀飲擇食，這是指兒童會吃水果，但只偏愛吃某一種類，例如只吃奇異果、香蕉，但不吃蘋果。雖然能攝取水果中的營養素，但長遠會導致飲食欠缺多元化。

挑食是較常出現的情況，家長應提供不同種類的食物，利用不同的食物形態，讓兒童對食物產生興趣，從而實踐多元化飲食。兒童往往較抗拒吃陌生的食物，家長需有耐性和心思，讓兒童多次重複接觸新食物。而食物的質感（如軟硬度）或味道也會影響兒童進食的次數，建議家長慢慢了解兒童的飲食習慣。家長亦可多與兒童正面地討論食物的顏色和味道，若兒童對食物產生興趣而作出新嘗試時，不妨適當地稱讚一下！（誰都愛被讚賞吧！）

| 4 人分量 |

三色水果雪葩 Tri-colour Fruity Sorbet

每份 (per serving)

熱量 (Energy)	碳水化合物 (Carbohydrate)	蛋白質 (Protein)	脂肪 (Fat)	膳食纖維 (Dietary Fibre)
63	16	1	0	1
千卡 (kcal)	克 (g)	克 (g)	克 (g)	克 (g)

蜜瓜雪葩材料

- ☐ 蜜瓜粒 1 杯
- ☐ 青檸汁 1 茶匙
- ☐ 楓糖漿 1 茶匙

哈密瓜雪葩材料

- ☐ 哈密瓜粒 1 杯
- ☐ 青檸汁 1 茶匙
- ☐ 楓糖漿 1 茶匙

西瓜雪葩材料

- ☐ 西瓜粒 1 杯
- ☐ 熟香蕉 1/2 隻
- ☐ 青檸汁 1 茶匙

做法

1. 蜜瓜粒、哈密瓜粒、西瓜粒分別放入密實袋，並置於冰箱冷藏 4 至 6 小時。
2. 將蜜瓜粒、青檸汁和楓糖漿放進攪拌機，攪拌至泥狀。
3. 將哈密瓜粒、青檸汁和楓糖漿放進攪拌機，攪拌至泥狀。
4. 將西瓜粒、香蕉、青檸汁放進攪拌機，攪拌至泥狀。
5. 將攪拌的果蓉分別放入三個密實袋內，放入冰箱冷藏 1 小時，取出，按壓約 2 分鐘後，再冷藏 1 小時，即可享用。

小助手參與步驟

營養小貼士

- 蜜瓜、哈密瓜、西瓜含有天然果糖和豐富水分，非常低脂，同時也含有多種營養素。
- 每杯淺綠色的蜜瓜粒（約 177 克）提供約 64 千卡熱量、1.4 克膳食纖維，還含豐富維他命 C、B_6、鉀質等。每杯橙黃色的哈密瓜（約 177 克）提供約 60 千卡熱量、1.6 克膳食纖維，還含有對眼睛和具抗氧化功效的 β- 胡蘿蔔素。紅彤彤的西瓜，每杯（約 154 克）提供約 47 千卡熱量和 0.6 克膳食纖維，含有維他命C、類胡蘿蔔素和茄紅素等抗氧化物，誰說「甜食沒好東西？」

選擇合適冰品消消暑

天氣酷熱，很多人難免想買杯雪糕消消暑。筆者不時會被問市面上多款雪糕、雪葩該如何選擇？冰凍甜點一般分為幾類，每款的製作方式和主要成分都不同：

- 傳統雪糕（Ice-cream）：一般以忌廉、牛奶、糖、蛋黃和其他配料混合而成。
- 意大利雪糕（Gelato）：相比傳統雪糕，一般使用較多牛奶和較少忌廉，乳脂含量相對較少。
- 雪葩（Sobert）：主要成分為水果蓉和糖，一般不含乳製品，適合對奶製品敏感或對乳糖不耐人士食用。
- 乳酪雪糕（Frozen Yogurt）：製作過程中多使用乳酪代替其他乳製品，相比傳統雪糕較低脂肪。

家長應留意產品的營養成分，會因應不同口味加入的配料和各品牌的配方而有所不同。想了解更多有關產品的成分和營養資料，最好細閱營養或食物標籤。

＊動動手！DIY 健康無添加糖冰凍甜品——將水果去皮、切塊，放入冰箱冷藏一晚，放入攪拌機打成雪葩。想令水果甜品口感更幼滑，可加入冰香蕉和少量低脂奶或乳酪。若覺得太麻煩，可直接將提子洗淨、切半或菠蘿切塊後放入冰箱冷藏，成為清甜冰涼的粒粒水果冰！

| 4 人分量 |

蘋果餡餅
Baked Apple Turnover

每份 (per serving)

熱量 (Energy)	碳水化合物 (Carbohydrate)	蛋白質 (Protein)	脂肪 (Fat)	膳食纖維 (Dietary Fibre)
119	19	5	2	2
千卡 (kcal)	克 (g)	克 (g)	克 (g)	克 (g)

材料

- ☐ 去皮方包 4 片（大，全麥麵包較佳）
- ☐ 雞蛋 1 隻
- ☐ 低脂鮮奶 1 湯匙

餡料

- ☐ 蘋果 80 克（去皮、切小粒）
- ☐ 黃糖 1 茶匙
- ☐ 開水 2 湯匙 - 1/4 杯
- ☐ 粟粉 1 茶匙
- ☐ 肉桂粉 1/4 茶匙
- ☐ 雲呢拿香油 1/2 茶匙

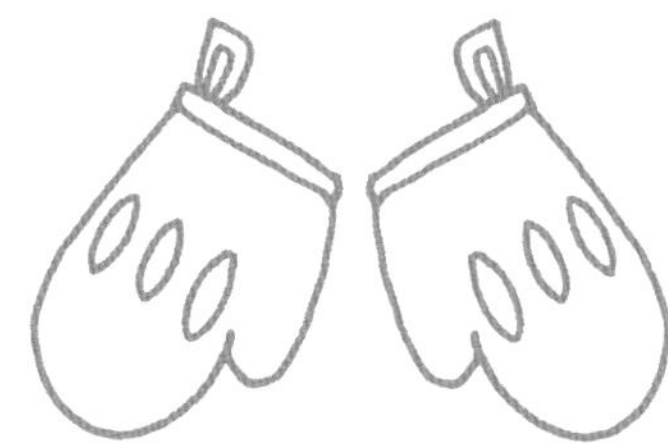

做法

1. 雞蛋和低脂鮮奶放入大碗，拌勻備用。
2. 除蘋果外，其他餡料放入小鍋，拌勻煮熱，加入蘋果粒用小火煮至變軟。
3. 方包用擀麵棒滾平，在其中一邊劃上 3 條斜線，於方包邊塗上適量蛋液，每件方包鋪上約 20 克蘋果餡。
4. 方包對摺，用叉子壓邊黏緊，於表面塗上一層薄薄的蛋液。
5. 放入已預熱 180°C焗爐，烤約 10 分鐘至表面金黃色（或放入氣炸鍋約 4 分鐘）即成。

小助手參與步驟

營養小貼士

- 坊間大部分批撻會用上酥皮作為主要食材，製作酥皮時，一般需要用上牛油或起酥油，其飽和油脂肪（甚至反式脂肪）含量較高，不利心臟健康。利用較低脂的方包代替酥皮，不但較健康，更讓製作過程變得簡單。
- 除了蘋果配上肉桂粉，以藍莓配乳酪、香蕉配花生醬也是不錯的餡料選擇。

輕怡甜點
| 2 人分量 |

夢幻 Smoothie

Dreamy Smoothie

第一層材料

- ☐ 低脂原味乳酪 1/4 杯
- ☐ 低脂鮮奶 1/4 杯
- ☐ 藍莓 1/2 杯
- ☐ 香蕉 1/2 隻
- ☐ 奇亞籽 1/2 湯匙

第二層材料

- ☐ 低脂原味乳酪 1/4 杯
- ☐ 低脂鮮奶 1/4 杯
- ☐ 芒果 1/2 杯
- ☐ 香蕉 1/2 隻
- ☐ 奇亞籽 1/2 湯匙

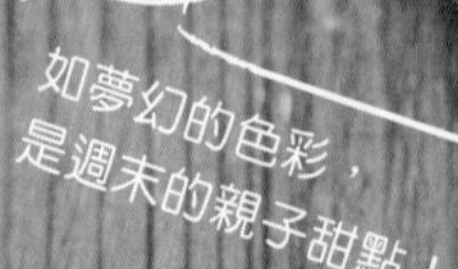
如夢幻的色彩，
是週末的親子甜點！

熱量 (Energy)	碳水化合物 (Carbohydrate)	蛋白質 (Protein)	脂肪 (Fat)	膳食纖維 (Dietary Fibre)
215	36	8	5	6
千卡 (kcal)	克 (g)	克 (g)	克 (g)	克 (g)

做法

1. 水果切片，放於冰箱冷藏，備用。
2. 將第一層所有材料倒入攪拌機攪拌，慢慢倒入杯內。
3. 攪拌機清理後，放入第二層所有材料攪拌，平均倒入杯內作第二層，即成！

小助手參與步驟

奇亞籽含豐富奧米加3脂肪酸、膳食纖維、蛋白質和多種礦物質，如鈣、鐵、鎂和鋅等，每15克奇亞籽提供約5克膳食纖維，屬高纖食品。

營養師聊天室

健康「果昔」方程式，你認識嗎？

「天氣咁熱，有咩飲品介紹？」很多人都會選擇珍珠奶茶、果茶和果汁飲品，而色彩繽紛的健康「果昔」（fresh fruit smoothies）屬健康之選！

「果昔」的營養含量均衡，提供的膳食纖維和蛋白質含量相比果汁和果茶高，更能製成多種款式，自由配搭一杯屬於自己的健康「果昔」！

低脂奶、低脂原味乳酪或無糖豆奶含豐富蛋白質，有助增加飽腹感。水果如藍莓、芒果、士多啤梨，再配上半隻香蕉讓「果昔」更香甜幼滑。選用冰鮮水果，效果更佳。另外，加入一杯綠葉蔬菜如菠菜或羽衣甘藍，熱量低之餘，更有助輕易達到每日攝取三份蔬菜的目標，可加入亞麻籽、奇亞籽和原味果仁等健康脂肪來源或含豐富膳食纖維的麥片，讓「果昔」變得更濃稠。如希望「果昔」增添風味，可加入肉桂粉或雲呢拿香油，將所有材料放進攪拌機打成奶昔即可。

口感軟綿，健康滋味！

輕怡甜點

| 4 人分量 |

雜果磚頭果凍 Mixed Fruit Jelly Brick

熱量 (Energy)	碳水化合物 (Carbohydrate)	蛋白質 (Protein)	脂肪 (Fat)	膳食纖維 (Dietary Fibre)
64	11	5	1	1
千卡 (kcal)	克 (g)	克 (g)	克 (g)	克 (g)

每份 (per serving)

材料

- ☐ 蜜瓜 150 克
- ☐ 哈密瓜 150 克
- ☐ 糖 15 克
- ☐ 清水 200 毫升
- ☐ 蝶豆花乾少許
- ☐ 魚膠粉 16 克
- ☐ 低脂鮮奶 100 毫升
- ☐ 薄荷葉少許（裝飾用）

做法

1. 將蜜瓜和哈密瓜挖成球狀，隨意混合後置於盤內。
2. 清水加入糖和適量蝶豆花乾，以小火加熱至糖溶化，製成淺藍色液體（如加入少量鮮檸檬汁可製成紫色）。
3. 熄火後，加入 10 克魚膠粉拌溶，放入已盛載水果的盤子中，冷藏至凝固。
4. 低脂鮮奶以小火加熱，加入 6 克魚膠粉拌勻，倒入已凝固的水果果凍表面。
5. 放於雪櫃冷藏至牛奶凝固，切塊，以薄荷葉裝飾即成。

小助手參與步驟

營養小貼士

- 加入蝶豆花乾製成不同顏色，增加兒童對食物和食物科學的興趣。
- 食譜可配搭兒童喜歡的水果製作代替現成果凍，減少游離糖的攝取。
- 新鮮水果當然是健康的選擇，但要留意避免使用新鮮菠蘿，因當中的酵素——菠蘿酶 (Bromelain) 會分解魚膠粉內的蛋白質，引致果凍未能成功凝固。
- 罐頭菠蘿經高溫處理，菠蘿酶的活性減低則可作代替品；要留意罐頭菠蘿的營養標籤，選擇較低糖分的款式。以果汁浸泡的罐頭水果，比糖漿浸泡的較低糖分，食譜使用的糖也可減掉。

仔仔經常傷風感冒，需要吃維他命 C 營養補充品嗎？

不少家長為兒童提供營養補充品，其中以維他命 C 最普遍。相信不少家長也聽過傷風感冒時要攝取維他命 C，但其實暫並沒有大型的統整科研證實維他命 C 能治療傷風感冒，攝取足夠的維他命 C 只能減低患上傷風感冒時的症狀。

雖然維他命 C 有助刺激抗體形成，從而協助增強免疫力，並協助上皮細胞的防禦功能，形成屏障抵抗病毒。只要從日常適量地進食帶酸性的水果，如橙、西柚、奇異果、木瓜等食物，也能從中攝取足夠的維他命 C。

以《中國居民膳食營養素參考攝入量（DRIS）》中，4 歲兒童維他命 C 的每天可耐受最高攝入量為 600 毫克，過量攝取會引致腹痛、腹瀉和增加患上腎結石的風險（千萬別以為量多就好）。由於坊間的維他命 C 營養補充品劑量各有不同，因此家長應避免兒童從營養補充品中攝取過多的維他命 C。當然，服用前應諮詢醫生和註冊營養師的意見。

食物	維他命 C（毫克）
甜椒（紅、生、1/2 杯）	95
橙（1 中型）	70
奇異果（1 中型）	64
甜椒（綠、生、1/2 杯）	60
西蘭花（熟、1/2 杯）	51
士多啤梨（新鮮、切粒、1/2 杯）	49
球芽甘藍（熟、1/2 杯）	48
西柚（1/2 中型）	39

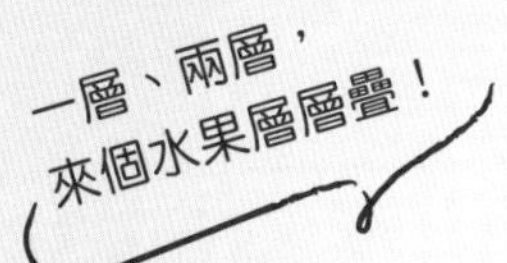

| 2 人分量 |

水果乳酪芭菲配奇亞籽果醬

Fruity Yogurt Parfait with Chia Seed Jam

每份 (per serving)

熱量 (Energy)	碳水化合物 (Carbohydrate)	蛋白質 (Protein)	脂肪 (Fat)	膳食纖維 (Dietary Fibre)
165 千卡 (kcal)	35 克 (g)	5 克 (g)	2 克 (g)	5 克 (g)

奇亞籽士多啤梨果醬材料

- ☐ 士多啤梨 80 克
- ☐ 蜜糖 2 克
- ☐ 奇亞籽 5 克
- ☐ 新鮮檸檬汁 3 克
- ☐ 新鮮檸檬皮適量

做法

❶ 士多啤梨洗淨，切粒，放入鍋內以細火邊攪拌邊煮至軟身。

❷ 將部分士多啤梨壓成蓉；保留部分果肉以增加口感。

❸ 試味後熄火，加入適量蜜糖、檸檬汁、檸檬皮及奇亞籽拌勻。

❹ 如喜歡，可額外加入奇亞籽並拌至喜愛的濃稠度，果醬冷卻後會變得更濃郁。

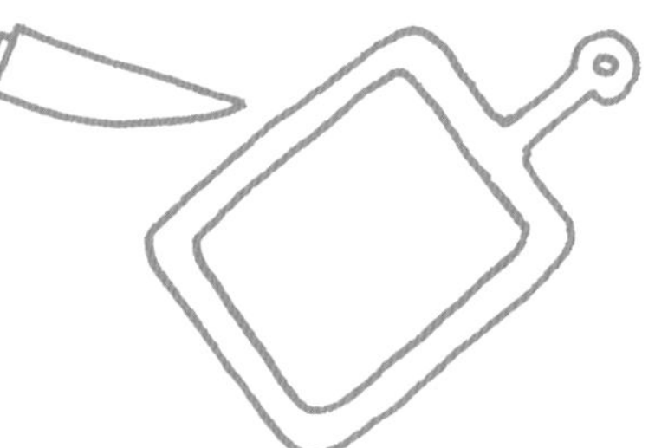

水果乳酪芭菲材料

- ☐ 低脂原味乳酪 1 小杯
- ☐ 奇異果 1 個
- ☐ 士多啤梨 2 粒
- ☐ 藍莓 1/2 盒
- ☐ 柑 1 個
- ☐ 黑莓 1/2 盒
- ☐ 粟米片 4 湯匙（低脂、低糖）
- ☐ 自製奇亞籽果醬 4 茶匙
- ☐ 薄荷葉適量（裝飾用）

做法

❶ 水果切粒或切片。

❷ 依序加入粟米片、低脂原味乳酪、生果及奇亞籽果醬，重複此步驟，最後加上薄荷葉裝飾即成。

- 新鮮水果及奇亞籽能提供膳食纖維，有助腸臟蠕動及預防便秘。
- 自製果醬可自行控制糖分的使用量，較坊間購買現成的果醬更健康。
- 低脂原味乳酪含有鈣質，有助骨骼健康。
- 不同顏色的水果顏色吸引，含多種植物化合物，具抗氧化功用。

營養師聊天室

乳酪大不同

市面上有不同的乳酪選擇，究竟乳酪（Yogurt）、希臘乳酪（Greek Yogurt）和希臘式乳酪（Greek-style Yogurt）有甚麼分別呢？作為家長應如何為兒童選擇合適的乳酪呢？

乳酪是利用加熱的牛奶，加入保加利亞乳桿菌和嗜熱鏈球菌經發酵製作，由於細菌將牛奶中的乳醣轉化為乳酸，因此原味乳酪帶有酸味和益生菌，患有乳醣不耐受症的人士可逐少淺嘗。一般的乳酪能提供蛋白質、鈣質和維他命 B_{12}。希臘乳酪製作原理相若，但多加了過濾的程序將乳清蛋白分離，因此質地較濃厚、酸味較突出。由於製作同等分量的希臘乳酪需使用更多牛奶，因此其蛋白質含量比一般乳酪高。一般希臘式乳酪只利用凝固劑模仿希臘乳酪的濃厚質感，因此其營養成分與一般乳酪分別不大。不同的產品配方不一，比較時建議多閱讀營養標籤為準。

無論任何一款乳酪，按以下條件選擇可作為健康小食：

- 低脂比全脂佳——乳酪以牛奶製作含有飽和脂肪；
- 原味比果味好——加入果肉同時亦有機會額外加入糖分；
- 成分表愈簡單愈好——表示產品沒有添加過多種類的防腐劑、人工色素或其他化學物。

| 2 人分量 |

芒果豆乳布丁
Mango Soy Milk Pudding

每份 (per serving)

熱量 (Energy)	碳水化合物 (Carbohydrate)	蛋白質 (Protein)	脂肪 (Fat)	膳食纖維 (Dietary Fibre)
73	7	5	3	1
千卡 (kcal)	克 (g)	克 (g)	克 (g)	克 (g)

材料

- ☐ 高鈣低糖豆漿 250 毫升
- ☐ 魚膠粉 3.5 克
- ☐ 芒果肉 1/4-1/2 杯

做法

1. 豆漿倒入鍋內，灑上魚膠粉待 5 分鐘，用小火將豆漿加熱，期間不停攪拌，直至魚膠粉完全溶解，熄火（不要煮滾豆漿）。
2. 以濾網隔走未溶解的魚膠粉及氣泡，倒進 2 個小杯，蓋好，放雪櫃冷藏約 6 小時或一晚製成布丁。
3. 芒果壓蓉，倒在布丁上，即可享用。

小助手參與步驟

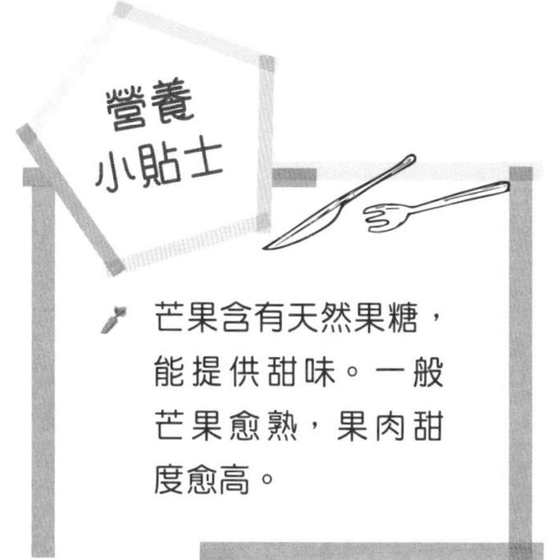

芒果含有天然果糖，能提供甜味。一般芒果愈熟，果肉甜度愈高。

市面上有多種植物奶，如何選擇？

牛奶為身體提供鈣質及蛋白質，建議兒童每天進食2份奶製品（1份約為240毫升）。市面上推出很多植物奶，營養成分跟牛奶有甚麼不一樣？這些產品可以代替牛奶嗎？

杏仁奶（Almond Milk）：碳水化合物及熱量較低，含不飽和脂肪和抗氧化物維他命E，但蛋白質和天然鈣含量較低。

穀物類植物奶：含較多碳水化合物，以米奶（Rice Milk）為例，每杯含 25 克（是牛奶的兩倍），蛋白質含量則較低。燕麥奶（Oat Milk）的蛋白質含量比米奶高，含有助降低壞膽固醇的水溶性纖維。

豆奶（Soy Milk）的蛋白質含量與牛奶相近，更含大豆異黃酮等抗氧化物，但黃豆的天然鈣量不高。

大家可根據身體狀況和喜好選購適合自己的奶類，由於每種飲品的配方不一，建議多閱讀營養標籤和成分表，以低糖或無添加糖的加鈣植物奶選擇為佳。

| 2 人分量 |

香蕉燕麥動物班戟

Banana Oat Animal Pancake

每份 (per serving)

＊不包括裝飾及配料 (excluding the toppings)

熱量 (Energy)	碳水化合物 (Carbohydrate)	蛋白質 (Protein)	脂肪 (Fat)	膳食纖維 (Dietary Fibre)
257	35	13	8	4
千卡 (kcal)	克 (g)	克 (g)	克 (g)	克 (g)

材料

- ☐ 熟香蕉 1 隻
- ☐ 雞蛋 1 隻
- ☐ 燕麥片 1/4 杯
- ☐ 泡打粉 1/8 茶匙
- ☐ 肉桂粉 1/4 茶匙
- ☐ 鹽 1/8 茶匙
- ☐ 芥花籽油 1 茶匙
- ☐ 低脂原味乳酪 1 杯
- ☐ 新鮮水果適量
- ☐ 非油炸原味果仁適量

做法

1. 香蕉、雞蛋、燕麥片、泡打粉、肉桂粉和鹽放入攪拌機，攪拌至軟滑。
2. 易潔鑊加熱下油，以廚房紙抹勻表面，加入 1/2 或 1/4 分量的班戟糊漿，將兩面煎至金黃色。
3. 放涼後，加上低脂原味乳酪、新鮮水果和果仁裝飾即成。（由於水果質地較軟，可安排兒童使用膠刀或曲奇餅模具，按喜好切出水果形狀裝飾。）

小助手參與步驟

營養小貼士

- 用攪拌機將所有材料混合能簡單製成班戟糊，代替現成的班戟預拌粉，可減少糖分的攝取。
- 班戟以燕麥片製作，增加膳食纖維含量，配合不同的水果裝飾成可愛圖案，吸引兒童食慾，並增加攝取新鮮水果。
- 燕麥含水溶性纖維、蛋白質、多種維他命及礦物質。以水果伴燕麥班戟，能攝取豐富之營養，而且賣相吸引。

喜慶節日食品

| 2 人分量 |

零失敗聖誕頭盤

Foolproof Christmas Appetizers

每份 (per serving)
* 不包括裝飾 (excluding the toppings)

熱量 (Energy)	碳水化合物 (Carbohydrate)	蛋白質 (Protein)	脂肪 (Fat)	膳食纖維 (Dietary Fibre)
122	19	4	4	2
千卡 (kcal)	克 (g)	克 (g)	克 (g)	克 (g)

材料

- ☐ 純麥比得包 1 片
- ☐ 牛油果醬 4 湯匙（做法參考 p.46）

裝飾

- ☐ 三色燈籠椒（切粒）
- ☐ 全麥早餐穀物條

做法

❶ 比得包切成 8 等份，為每一小塊塗上約 1-1.5 茶匙牛油果醬。

❷ 用廚房紙吸乾燈籠椒的多餘水分。

❸ 以燈籠椒粒為聖誕裝飾；全麥早餐穀物條為樹幹，為小聖誕樹裝飾。

小助手參與步驟

營養小貼士

- 純麥比得包用途廣泛，除了可配上牛油果醬和鷹嘴豆泥當小食外，還可以當作 DIY 薄餅的餅底，方便又健康。
- 不妨利用純麥比得包製作三文治，切半後打開袋口放入餡料，方便兒童食用。
- 除燈籠椒粒，亦可以切片車厘茄或其他新鮮蔬果作裝飾。
- 全麥早餐穀物條加入低脂或脫脂奶、無添加糖的水果乾，可製作簡易早餐。選擇早餐穀物片時需閱讀營養標籤，以較低糖的選擇為佳。

|10個/人分量|

新春開心能量球

The Perfect Energy Balls for Chinese New Year

每份 (per serving)

熱量 (Energy)	碳水化合物 (Carbohydrate)	蛋白質 (Protein)	脂肪 (Fat)	膳食纖維 (Dietary Fibre)
69	8	3	3	2
千卡 (kcal)	克 (g)	克 (g)	克 (g)	克 (g)

材料

- ☐ 原片燕麥 4 湯匙
- ☐ 純花生醬 2 湯匙
- ☐ 開心果 1 湯匙
- ☐ 杏仁 1 湯匙
- ☐ 紅莓乾或無糖提子乾 1 湯匙
- ☐ 杞子 1 湯匙
- ☐ 蜜糖 1 茶匙
- ☐ 肉桂粉 1 茶匙（隨意）

能量球造型

做法

1. 開心果、杏仁、紅莓乾、杞子略為切碎，保留少許原粒裝飾用。
2. 大碗內將原片燕麥、開心果、杏仁、肉桂粉混合，加入純花生醬、蜜糖、紅莓乾及杞子拌勻。
3. 保鮮紙上加入 1 湯匙混合好的材料，包成球體即成。

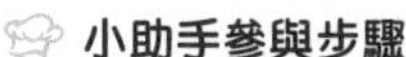

小助手參與步驟

營養小貼士

- 燕麥含水溶性纖維，有助降低體內膽固醇及血糖水平。
- 純花生醬沒有添加任何油、糖及鹽，是較為健康的選擇。
- 原味果仁含有不飽和脂肪、纖維及蛋白質，可當作平日的健康小食。
- 利用肉桂粉帶出甜味，卻不會增加糖分攝取。

親子得意賀年小吃，
為新年添氣氛！
喜慶節日食品

| 12 件 / 人分量 |

大吉大利糕點
Pumpkin Chinese New Year Treats

每份 (per serving)
* 不包括裝飾 (excluding the toppings)

熱量 (Energy)	碳水化合物 (Carbohydrate)	蛋白質 (Protein)	脂肪 (Fat)	膳食纖維 (Dietary Fibre)
49	9	1	1	1
千卡 (kcal)	克 (g)	克 (g)	克 (g)	克 (g)

材料

- ☐ 南瓜 150 克（去皮、去籽，切粒）
- ☐ 蔗糖 15 克
- ☐ 鹽 1/8 茶匙
- ☐ 糯米粉 75 克

餡料

- ☐ 熟香蕉 25 克
- ☐ 花生碎 25 克
- ☐ 蔗糖 15 克

裝飾

- ☐ 南瓜籽、高纖早餐穀物條各適量

做法

1. 南瓜隔水蒸至軟身，如需要可去掉多餘水分。
2. 南瓜用叉子壓成蓉，拌入蔗糖、鹽和糯米粉搓成麵糰。
3. 香蕉壓成蓉，加入花生碎和蔗糖碎成餡料。
4. 麵糰平均分成 12 份，搓成小球用手壓扁，包入 1 茶匙餡料搓成球狀，用牙籤在表面輕戳小孔。
5. 麵糰用中火蒸約 10-12 分鐘，加上南瓜籽和高纖早餐穀物條裝飾，即可享用。

小助手參與步驟

大吉大利糕點造型

營養小貼士

- 南瓜含有類胡蘿蔔素、膳食纖維、多種維他命和礦物質。
- 南瓜籽含有豐富營養，與其他果仁一樣含有不飽和脂肪、高抗氧化營養素如鋅質、鐵質，更含有助心臟和骨骼健康的鎂質。
- 食譜利用熟香蕉製作餡料，有助提供甜味，更能增加膳食纖維。

小狗造型湯圓，
趣味滿分！
喜慶節日食品
| 5 人分量 |

柴犬湯圓

Shiba Inu Glutinous Rice Balls

每份 (per serving)

熱量 (Energy)	碳水化合物 (Carbohydrate)	蛋白質 (Protein)	脂肪 (Fat)	膳食纖維 (Dietary Fibre)
108	20	4	2	1
千卡 (kcal)	克 (g)	克 (g)	克 (g)	克 (g)

材料

- ☐ 糯米粉 90 克
- ☐ 豆腐 100 克
- ☐ 生抽 2 湯匙
- ☐ 黑糖 15 克
- ☐ 清水 50 毫升
- ☐ 粟粉 1 茶匙（以少許水調勻）
- ☐ 黑芝麻適量

做法

1. 糯米粉與豆腐混合，取一部分（用作耳朵、眼眉、鼻子和尾巴），餘下的分為 10 份，搓成球狀。
2. 將預先取出的部分製成 5 對眼眉、5 對耳朵、5 個鼻子和 5 條尾巴，並與球狀合併，用少許水黏實，製成柴犬的頭部和臀部。
3. 放進沸水煮至浮起，盛起備用。
4. 於小鍋內加入生抽、黑糖和清水，以小火煮至糖溶解，加入粟粉水煮至黏稠。
5. 將醬汁輕掃於柴犬形成毛髮，並加上黑芝麻裝飾即成。

小助手參與步驟

營養小貼士

- 用豆腐製作湯圓，與一般湯圓相比增加了蛋白質含量，並減少糖分的使用。
- 醬汁的生抽和糖分可按個人喜好調節，有利進一步減少糖分和鈉質的攝取。
- 坊間購買的湯圓大多使用了棕櫚油製作，棕櫚油含較高的飽和脂肪，過量攝取會增加患上心血管疾病的風險。

健康素食選擇

營養師經常鼓勵大眾均衡飲食，是否不主張兒童跟從素食飲食呢？

若因宗教信仰或信念（如保護環境或為了維護動物權益）而進行素食，筆者當然尊重並教導如何從素食攝取足夠的營養素，以免影響成長。

素食泛指以植物為主要食糧，同時不進食部分或全部動物食物或其製品的飲食模式，可分為以下四大類：

- 蛋奶素食 (Lacto-ovo-vegetarian)：會進食蛋類和奶類製品。
- 蛋素食 (Ovo-vegetarian)：會進食蛋類食物。
- 奶素食 (Lacto-vegetarian)：會進食奶類食物。
- 全素食 (Vegan)：只進食植物性食物，即使來自動物的食品（如蜜糖和燕窩）都不吃。

進食素食帶來的益處：

- 食物大多屬高纖、低脂和低熱量食物，且不含膽固醇，能增加飽肚感，亦有助預防便秘、減低膽固醇過高和體重上升的機會。
- 蔬果和豆類含有豐富的抗氧化物、植物化學物維他命和礦物質，適量進食能增強抵抗力，有助減低患上高血壓、心臟病和糖尿病等慢性疾病的機會。

素食並不代表完全健康，有以下原因：

- 應留意一些經油炸製作的豆類製品，如素鴨、炸豆泡、炸枝竹，加入醬汁製作的素菜等。
- 近年出現一些「仿肉類素食品」，製作過程中加入額外油分和鹽分，過量進食會引致體重上升，從而增加患上慢性疾病的風險。

素食者要均衡進食不同種類的食物，同時留意以免缺乏以下 5 種營養素！

1. 蛋白質：大多植物性食物雖含有蛋白質，但其質素並未如肉類及其代替品和奶類，因此素食者須進食足夠和不同種類的種籽果仁、乾豆類、大豆製品和全穀物，以確保營養均衡。
2. 維他命 B_{12}：主要來自肉、蛋和奶類食品，全素食者應食用添加了維生素 B_{12} 的豆奶和早餐穀物片，同時建議向醫生或註冊營養師查詢使用補充品的需要。
3. 維他命 D：適量曬太陽使皮膚製造維他命 D，幫助身體吸收鈣質，預防骨質流失。
4. 鈣質：透過進食硬豆腐、加鈣豆漿、種籽果仁和深綠色蔬菜如西蘭花、菜心、芥蘭等補充。
5. 鐵質：進食菠菜、乾豆類、黑木耳、提子乾、早餐穀物片和全麥麵包等攝取鐵質，同時進食含豐富維他命 C 食物，如番茄、西蘭花和橙等，有助鐵質吸收。

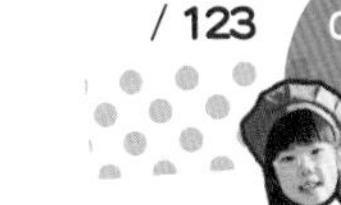

Contents

Teddy Bear Steamed Banana Cupcakes

| makes 6 servings | **refer to p.38 |**

Ingredients

75 g self-raising flour / 1 whole egg / 1 tbsp cane sugar / 1 tbsp vegetable oil / 25 g low fat milk / 1 medium banana (very ripe, mashed)

For decoration

raisins (with no added sugar) / white chocolate buttons / pumpkin seeds / chocolate (melted) or all-natural peanut butter

Method

1. Prepare a steamer, bring water to a boil.
2. In a large mixing bowl, whisk the egg until foamy and then add the oil, sugar and low fat milk, mix for another 5 minutes, until the mixture turns thick and pale yellow. Add in the mashed banana, mix well. 🧑‍🍳
3. Sift in the self-raising flour slowly, mix well until well incorporated. Try not to over mix the batter. Pour the mixture into 6 cupcake molds (90% full). 🧑‍🍳
4. Steam the cakes for around 15 minutes or until cooked.
5. Decorate the cupcakes with the raisins, white chocolate buttons, pumpkin seeds and melted chocolate. Serve. 🧑‍🍳

🧑‍🍳 Kid-friendly steps

Nutrition tips

- Mashed ripe banana offers extra fibre and nutrients to the recipe. Its natural sweetness and texture can be used to replace some of the fat and sugar in the recipe.
- Vegetable oil is used in place of butter to limit unhealthy saturated fat intake.
- Choose nuts, seeds and dried fruits with no added sugar or salt whenever possible.
- Let your kids try out and pick their favourite nuts, seeds and dried fruits for the recipe! The more they get to explore, the more interest they have in food!

Butterfly / Bowtie Bun

| makes 8 servings | refer to p.40 |

Ingredients

250 g bread flour / 25 g sugar / 2 g salt / 3 g yeast / 1 whole egg / 100 g low fat milk / 25 g canola oil / a small amount of bread flour (for decoration)

Method

1. Combine bread flour, sugar, salt and yeast into a big bowl. (Take note that yeast should be added separately from sugar and salt to prevent it from spoiling.)
2. Add egg and low fat milk, mix the dough until it is not sticky. Transfer the dough onto a clean surface and knead until smooth and elastic. (For kneading: press downwards onto the dough with your palm, then push it upwards, fold it back to the original position and repeat this step.)
3. Take out a small portion of the dough, stretch it with your fingers to form a hole.
4. If the hole has smooth edges, place it back to the rest of the dough and add in the canola oil slowly. Keep kneading until the dough fully absorbs the oil (The dough will become slightly sticky after adding the oil, but don't worry and keep kneading.)
5. Roll the dough into a ball and place it into a bowl, cover and let it rise for an hour, or until the dough doubles in size.
6. Place the fermented dough on the table and punch it down with your hands to release air. Divide it into 8 equal portions of around 50 g each, rolling each portion into a ball.
7. Cover the balls of dough with cling film and leave to rest for 20 minutes. Use a rolling pin to flatten the balls into round disks, then cut them according to the pictures with a knife.
8. Place the butterfly or bowtie bread doughs onto a baking sheet lined with baking paper, then cover them with cling film. Let them rest for around 40 minutes.
9. Sprinkle a small amount of flour onto the rolls and bake them in a preheated oven at 180°C for 15 minutes. Let cool and serve!

Kid-friendly steps (steps 2, 4, 6, 7)

Nutrition tips

- Eggs and low fat milk offer extra quality protein to children.
- These cute buns are perfectly portioned and easy to serve, children will be delighted to have them regularly.
- Each serving provides carbohydrates equivalent to half a bowl of rice. Serve them along with one serving of low fat dairy product (such as 2 slices of low fat cheese or 1 cup of semi-skimmed milk) for a more balanced breakfast option.
- You may replace 1/3 of bread flour with oat bread flour or oat flour, to increase your daily dietary fibre intake.

Pumpkin Bread with Chestnut Filling

| makes 8 servings | refer to p.43 |

Ingredients

100 g pumpkin / 210 g bread flour / 50 g low fat milk / 25 g sugar / 2 g salt / 3 g yeast / 20 g canola oil / 8 pieces chestnuts (ready-to-eat) / 8 pieces pumpkin seeds

Method

1. Remove the skin and seeds of the pumpkin and then slice into small pieces. Steam over medium heat for 15 minutes until tender. Mash the pumpkin and set aside to cool.
2. Combine the bread flour, mashed pumpkin, sugar, salt and yeast in a large bowl. Check the texture of the dough and pour in low fat milk a little at a time since the water content of steamed pumpkin varies each time.
3. Knead the dough until it does not stick to your hands or the bowl. Transfer the dough onto a clean surface and knead until smooth and elastic. (For kneading: press downwards onto the dough with your palm, then push it upwards, fold it back to the original position and repeat this step.)
4. Take out a small portion of the dough, stretch it with your fingers to form a hole. If the hole has smooth edges, place it back to the rest of the dough and add in the canola oil slowly. Keep kneading until the dough fully absorbs the oil (the dough will become slightly sticky again when adding oil, continue to knead the dough).
5. Roll the dough into a round ball and place it into a bowl, cover and let it rise for an hour, or until the dough doubles in size.
6. Place the fermented dough on the table and punch it down with your hands to release air. Divide it into 8 equal portions of around 50 g each, shape the dough pieces into balls. Cover them with cling film and leave the dough to rest for 20 minutes. Use a rolling pin to flatten the dough into round disks, place the chestnut in the center and wrap up.
7. Dust your knife or toothpick with some flour. Score the dough as shown, and add pumpkin seeds on top for decoration.
8. Place the dough onto a baking sheet lined with baking paper and cover it with cling film. Let them rise for 40 minutes.
9. Bake them in a preheated oven at 180°C for 15 minutes, let cool and serve!

Kid-friendly steps

Nutrition tips

- Pumpkin adds vibrant colour and nutrients such as beta carotene and dietary fibre to the recipe.
- Your body converts beta-carotene to vitamin A, a nutrient that is great for our eyes and can help prevent night blindness. Beta carotene is a fat soluble vitamin, however, high amounts of beta-carotene can turn your skin yellow or orange. The skin colour will return to normal once you reduce your intake of beta-carotene.

Your Everyday Avocado Spread

| makes 4 servings | refer to p.46 |

Ingredients

1 avocado (medium-sized) / 2 tbsp diced tomatoes / 2 tbsp diced onion / 1 tsp grated garlic / 1 tsp olive oil / 1/4 lemon (juiced) / salt / ground black pepper

Method

1. In a small nonstick pan over medium-low heat, heat oil, add in the tomato, onion, and garlic, let it cook for 2-3 minutes.
2. Cut the avocado in half, remove the pit, and scoop out the flesh.
3. In a large bowl, mix in all the ingredients, mash the avocado to desired texture, season it with lemon juice, salt and pepper, enjoy!

Kid-friendly steps

Nutrition tips

- Avocados are a great source of dietary fibre, potassium, and antioxidants such as vitamin E. It is also rich in unsaturated fatty acids, which could help keep your heart healthy.
- Half an avocado at around 100 g contains 15 g of fat (which is equivalent to around 3 tsp of oil), though it is made up of mostly healthy fat - unsaturated fat, we should still be cautious with the serving size.
- Diced onion and tomato provides extra flavour, texture and nutrients to the avocado spread.
- Lemon is rich in vitamin C and helps prevent the avocado from browning.

Curly Kale Crunch

| makes 2 servings | **refer to p.49 |**

Ingredients

2 cups curly kale leaves (washed and dried) / 2 tsp olive oil

Seasoning

1/8-1/4 tsp salt / 1/4 lime (juiced)

Method

1. Preheat the oven to 175°C.
2. Put all the ingredients in a large bowl and mix until curly kale is coated with oil. 🧑‍🍳
3. Spread kale into a single layer on a lined baking sheet, bake for around 10 minutes, watch them closely in the oven make sure they are crispy and not burnt. (Rotate the pan midway through the bake), season with salt and lime juice as desired.

🧑‍🍳 Kid-friendly steps

Nutrition tips

- Curly kale has always been hailed as a superfood as it is rich in nutrients such as vitamin A, vitamin C, vitamin K, lutein, zeaxanthin, calcium, potassium and dietary fibre.
- Kale is very high in nutrients and low in calories, making it one of the most nutrient-dense foods on the planet.
- Curly Kale Crunch are a great healthy substitute for potato chips that are usually higher in calories, fat and salt content.

Hong Kong Style French Toast

| makes 2 servings | refer to p.52 |

Ingredients

2 slices sandwich bread (large, preferably whole wheat) / 2 tsp-1 tbsp all-natural peanut butter / 1/2 medium banana (very ripe, mashed) / 1 whole egg / 2 tbsp low fat milk / 1 tsp vegetable oil / 1/2 cup blueberries and strawberries / maple syrup or honey (optional)

Method

1. Whisk together low fat milk and eggs, pour it into a shallow container. 🍳
2. Remove the crust and evenly spread the peanut butter and banana puree on one side of the bread, sandwich the two bread slices, cut it into 4 small pieces or other shapes as desired and dip the bread into the egg mixture. 🍳
3. In a small nonstick pan over medium-low heat, heat oil and add the soaked bread. Fry until golden brown on both sides.
4. Serve with berries and a small amount of syrup. 🍳

🍳 Kid-friendly steps

Nutrition tips

- Instead of deep frying, this recipe uses a small amount of oil to pan fry the toast to reduce fat intake.
- Mashed ripe banana and all-natural peanut butter are both more nutritious substitutes for usual sandwich fillings such as condensed milk and cream.
- Berries give the dish extra colour, sweetness and a boost of nutrients.

Shake Shake Baked Potato Wedges

| makes 2 servings | refer to p.54 |

Ingredients

200 g potato / 1 tsp garlic powder (unsalted) / 1/4 tsp salt / 1 tsp paprika powder / 1 tbsp parmesan cheese / 2 tsp vegetable oil

Creamy yogurt dip

1/4 cup low fat plain Greek yogurt
1-2 tsp lemon juice
1 tsp garlic (finely minced)
1 tbsp chives (finely minced) / 1/8 tsp salt
1/8 tsp ground black pepper

Method

1. Rinse the potatoes, cut into small wedges and soak in warm water for 15 minutes. Preheat the oven to 200°C.
2. Line your baking sheet with baking paper and set aside. For the creamy yogurt dip, simply combine all ingredients, and let it sit in the fridge for at least 15 minutes.
3. Wash away the surface starch of the potato wedges, pat dry to remove excess water.
4. Place the potato wedges into a container. Add garlic powder, salt, paprika powder, parmesan cheese and vegetable oil, then SHAKE!
5. Spread the potatoes evenly on the lined baking sheet and bake at 200°C for 30-35 minutes or until golden brown and fully cooked (for a more even texture, flip the wedges midway).
6. Serve with the creamy yogurt dip.

Kid-friendly step

Nutrition tips

- This recipe is a healthier alternative to deep-fried fries, as we have more control over the use of oil and seasoning.
- Garlic powder and paprika are used to provide flavours to the wedges in place of salt.
- We can also bake other root vegetables such as sweet potatoes and carrot. Root vegetables are highly nutritious especially when we keep the skin on, and could be a good substitute for bread and rice occasionally.

Cheesy Rolled Omelette

| makes 3 serving | refer to p.56 |

Ingredients

3 whole eggs / 2 tbsp finely diced onion / 3 tbsp finely diced carrot / 3 tbsp finely diced broccoli / 1 slice low fat cheese / salt / ground black pepper / 1-2 tsp vegetable oil

Method

1. In a large bowl, whisk the eggs and mix in other ingredients (except cheese) until well incorporated. 🍳
2. Preheat a non-stick pan and evenly coat with oil. Pour 1/3 cup of the egg mixture, tilt the pan left and right, try to spread the egg mixture evenly over the entire surface of the pan.
3. Cook the egg mixture over low-medium heat until it's half-cooked, add the cheese slices evenly, and start to roll the egg sheet into egg rolls. After the first layer is completed, apply a small amount of oil on the surface of the pan again and pour in another 1/3 cup of egg mixture (the egg mixture must be connected to the rolled egg roll).
4. Repeat the steps until all the egg mixture is used up. Once the egg roll is fully cooked, let it cool for a bit and you may shape and slice them up into small pieces, serve.

🍳 Kid-friendly steps

Nutrition tips

- Eggs are not only affordable and easily available, but are also a source of high-quality protein. In addition to boiling, steaming, and stir-frying, we can also add in different ingredients to prepare a variety of dishes! Yet, be sure to cook the eggs thoroughly until the yolk and egg whites become firm to reduce the chance of causing food-borne illness.
- Apart from low fat cheese, you can also add in other healthy ingredients such as seaweed, tuna soaked in water, corn kernels, etc. to make a variety of flavors as desired.

Baked Crispy Chicken Nuggets

| makes 4 servings | refer to p.58 |

Ingredients

200 g chicken breast (cut into bite-size pieces, 20 g each) / 2 tbsp skimmed plain yogurt / 1/2-1 cup plain corn flakes (crushed) / 2 tbsp parmesan cheese / 1/2 tsp garlic powder (unsalted) / 1/8 tsp salt / ground black pepper / olive oil (cooking spray)

Method

1. Preheat the oven to 200°C. Line the baking tray with parchment paper and set aside.
2. Marinate the chicken with salt, black pepper and yogurt in the refrigerator for at least 15 minutes.
3. In a large bowl, add crushed corn flakes, grated parmesan cheese, garlic powder, salt, and black pepper, mix well.
4. Coat the marinated chicken with corn flakes and seasoning. Evenly spray some olive oil on them (optional).
5. Let it bake for about 15-18 minutes, or until the chicken is cooked through and until the desired crispiness is achieved (for a more even texture, flip the nuggets midway).

Kid-friendly steps

Nutrition tips

- Chicken nuggets are usually deep-fried and can be quite high in fat. This recipe is a healthier alternative, using lean chicken breasts as the main ingredient and baking them instead.
- Plain yogurt can not only be eaten directly with fruits and nuts but can also be used to make low fat sauce or used as a marinade to tenderize meat.

Japanese Cheesy Salmon Cubes

| makes 4 servings | refer to p.60 |

Ingredients

1 potato (medium-sized) / 150 g salmon fillets / 1 whole egg / 2 slices high calcium low fat cheese / 1/4 tsp salt / 2 tsp dill / 10 g flour / 2 tsp olive oil

Method

1. Rinse and peel the potatoes. Dice them into cubes and place in boiling water, cook them until soft. Strain and set aside.
2. Marinate the salmon with salt and dill for about 15 minutes, set aside.
3. Add 1 tsp of olive oil into a non-stick pan over medium heat. Add in the salmon and let it cook through. Break them into small pieces with a fork and set aside.
4. Mash the potatoes in a bowl. Mix in the egg and salmon then mix well.
5. Divide the salmon mixture into 16 portions. Flatten it in your hand and place some cheese onto the center. Form into a cube shape. Sprinkle with flour.
6. Add the remaining 1 tsp of olive oil into the non-stick pan with medium heat. Pan fry the salmon cubes until each side is golden brown. Serve.

Kid-friendly steps

Nutrition tips

- Salmon and egg are both good sources of protein, aiding in cell and muscle repair, as well as forming white blood cells and antibodies to boost our immune system.
- Salmon is rich in the polyunsaturated fats--Omega-3 fatty acids, a type of fat that the body cannot make on its own.
- The American Heart Association recommends eating at least 2 servings of fish (particularly fatty fish) per week. Studies show that consuming fatty fish rich in omega-3 fatty acids can help reduce stress and anxiety.
- If you dislike fish or are vegetarian/vegan, chia seeds, flaxseed or walnuts are good sources of Omega-3 fatty acids.

Four Seasons Shao Mai

| makes 10 pieces/servings | refer to p.62 |

Ingredients

10 pieces squared dumpling wrappers / 100 g minced lean pork / 2 tbsp black fungus / 2 tbsp carrot / 2 tbsp corn kernels / 2 tbsp peas / 2 tsp soy sauce / 1/3 tsp sesame oil / 1/4 tsp grated ginger

Method

1. Add grated ginger, soy sauce and sesame oil to the minced pork and marinate for 15 minutes.
2. Boil the black fungus, carrot, corn kernels and peas. Let them cool and dice separately.
3. Place a heaping teaspoon of pork onto the center of the dumpling wrapper. Fold the wrapper, add a bit of water on the edges to help stick them together.
4. Fold the short edge to the point of the centre and press well. Slowly open the 4 pockets on the side.
5. Fill each pocket with diced vegetables as shown (1/2 tsp each).
6. Bring the water to a boil. Steam the Shao Mai for 10 minutes until completely cooked and serve.

Kid-friendly steps

Nutrition tips

- The Four Seasons Shao Mai is healthier than regular Shao Mai in the market. Not only is it made of fresh pork, but the addition of various vegetables increase the dietary fibre content.
- Steaming can minimize additional oil used in cooking, thus is a healthier cooking method.
- Depending on the amount of minced pork mixture, you may adjust the size and portion of the Shao Mai, it could be served as a main dish or snack.

Animal Garden Spinach Soup

| makes 2 servings | refer to p.64 |

Ingredients

2 slices bread / 250 g spinach / 1/2 potato (medium-sized) / 1/2 onion (diced) / 1 clove garlic (minced) / 300ml low fat milk / 250ml low sodium chicken stock / fresh lemon juice (1/4 lemon) / water / 1 tbsp olive oil

For decoration

15ml low fat milk

Method

1. Cut the bread into various animal shapes with a cookie cutter. Toast the bread and set aside. 🍳
2. Rinse and peel the potatoes. Cut into small pieces and set aside.
3. Heat up the oil in a non-stick pan over low heat and stir-fry the onion and garlic until fragrant. Add the potato pieces and fry slightly. Add the chicken stock and cook for 10 minutes until the potatoes soften.
4. Add the low fat milk, lemon juice and half of the spinach. Let it simmer for around 15 minutes. Set aside to cool for 5 minutes.
5. Pour soup into a blender and blend until smooth, then add the remaining spinach and blend until smooth. 🍳
6. Heat the soup in the pot (add some water if the soup is too thick and creamy).
7. Pour the soup into bowls, then serve with the toasts and a drizzle of low fat milk. 🍳

🍳 Kid-friendly steps

Nutrition tips

- Western soups are usually more nutritionally dense than Chinese clear broth since all the ingredients are blended with the stock and consumed.
- Spinach is rich in iron and folic acid, containing around 2.7mg of iron and 194μg of folic acid. These aid in the production of cells and red blood cells.
- This recipe uses potato to thicken the soup, this could help lower the overall fat content of the soup while adding more flavours.

Build-your-own Taiwanese Egg Crepe

| makes 3 servings | refer to p.66 |

Ingredients

3 whole eggs

Batter

40 g all-purpose flour / 40 g oat flour / 20 g cornflour / 200 g drinking water / spring onion (chopped) / salt / pepper / 1-2 tsp oil

Sauce

1 tbsp reduced sodium soy sauce / 1 tsp minced garlic / 1 tsp sesame oil / 1/2 tbsp drinking water

Fillings

canned tuna in water / low fat cheese / corn kernel / cooked shredded chicken / cooked mushroom slices (optional)

Method

1. Add the batter ingredients into a large bowl and mix well, set aside. In another bowl, combine all ingredients for the sauce and mix well.
2. Now heat the oil in a small non-stick pan. Pour in 1/3 cup of batter and quickly swirl until it reaches the edges. Cook for 1 minute until the crepe starts to cook.
3. Once the crepe is cooked, add an egg, spread it across the top in a thin layer. Continue to cook until the egg starts to set, add the filling of your choice, roll up the crepe, and coat it with sauce. Your egg crepe is ready to serve!

Kid-friendly steps

Nutrition tips

- To make the crepe more nutritious and with higher dietary fibre content, we have replaced some regular flour with oat flour. You may purchase oat flour or make oat flour by simply blending oats in a food processor or blender.
- For the fillings, try to avoid the use of processed meat or pickled vegetables. Choose fresh ingredients whenever possible. If needed choose canned tuna soaked in water instead of those soaked in oil.
- Add in a variety of vegetables to make the dish more colourful and attractive.
- Using natural seasonings such as garlic may help enhance the flavour of dishes and reduce the use of high sodium ready-made sauces. Read the nutrition labels of premade sauces, and pick the ones with the lower sodium content.

Shrimp Rice Balls

| makes 4 servings | refer to p.68 |

Ingredients

4 shelled shrimps / 2/3 bowl cooked rice (with oat) / 1 tbsp sushi seasoning / 1 tbsp corn kernels / 1 tbsp diced carrot / 2 tbsp shredded sushi seaweed / 1/2 tsp vegetable oil / 1/8 tsp salt / pepper

Method

1. Marinate the shrimps with salt and pepper for 15 minutes, then steam until fully cooked.
2. Heat up a non-stick pan and add 1/2 tsp of vegetable oil to stir-fry the carrot. until done.
3. Combine oat rice, sushi seasoning, corn kernels, carrot and chopped seaweed in a bowl, mix well.
4. Prepare a piece of cling wrap, place the shrimp in the center, then spread a layer of rice mixture on top, shape it into a ball, squeeze well and serve.

Kid-friendly steps

Nutrition tips

- Replacing part of the white rice with oats for rice and adding in all sorts of vegetables can not only increase the dietary fibre content of the dish, but also enhance the flavour profile and colour of the dish.
- Shrimp is a low-fat, lower-calorie source of protein.
- Sashimi is commonly used in Japanese dishes. However uncooked food especially meat has the potential to carry food-borne pathogens that can cause illness, therefore it is not recommended for children to consume.
- You may design your rice ball according to your preferences. For example, you may replace the shrimp from the recipe with cooked chicken or salmon!

Pan-fried Tofu Lotus Root Patty

| makes 5 servings | refer to p.70 |

Ingredients

200 g firm tofu / 150 g minced chicken / 1 whole egg / 2 tbsp diced carrot / 2 tbsp diced onion / 20 g breadcrumbs / 10 slices lotus root / 2 tsp canola oil / some edamame (for decoration)

Seasoning

1/8 tsp salt / 1/2 tsp sugar / 1 tsp soy sauce / 1 tsp sesame oil

Method

1. Pat the firm tofu dry with paper towels to remove excess water.
2. Mash the firm tofu and mix in minced chicken, egg, diced carrot and onion, breadcrumbs and seasoning. Mix well and set aside.
3. Heat up some oil on a non-stick pan. Divide the tofu chicken mixture into 10 round patties.
4. Cook them over medium heat until half cooked, then place lotus root slices onto each patty, add in some edamame as decoration. Flip over and let it cook until lightly brown, and thoroughly cooked. Serve.

Kid-friendly steps

Nutrition tips

- Tofu and chicken are soft and easier to chew, lotus root slices are harder. Parents may adjust the thickness of the lotus root according to children's chewing ability.
- Lotus root offers dietary fibre, providing 5 g per 100 g. Let children learn more about this vegetable by making and tasting this dish.
- Children can train muscle strength of their hands and fingers through cooking.

Japanese Steamed Egg

| makes 4 servings | refer to p.72 |

Ingredients

120 g frozen chicken thigh / 4 pieces frozen shrimps / 4 pieces fresh mushroom (small) / 2 okras / 8 slices carrot / 2 whole eggs / 1/4 tsp soy sauce (to marinade the chicken) / 300 ml hot water / 1 tsp katsuo dashi

Seasoning

1/4 tsp salt / 1/2 tsp sugar / 1 tsp soy sauce

Method

1. Rinse and trim off the chicken skin and fat. Pat dry with a kitchen towel and cut into small pieces; add soy sauce and marinate for around 15 minutes, set aside.
2. Rinse the frozen shrimps and pat dry with a kitchen towel, set aside.
3. Rub and clean the mushrooms and cut a flower pattern on the mushroom with a small knife. Cut the okra into star shapes. Slice the carrot, then cut out different shapes using cookie cutters. Cook the carrot and okra in boiling water until softened, then set aside to cool.
4. Mix katsuo dashi with hot water, add seasoning and set aside to cool. Add in the egg and mix well.
5. In a container, add chicken and shrimp, pass the egg mixture through a sieve, pour the mixture evenly in 2 small jars. Lastly, add fresh mushroom, carrot slices and okra on top.
6. Wrap the surface loosely with a cling film. Steam over low heat for around 10 minutes, serve.

Kid-friendly steps

Nutrition tips

- Frozen skinless chicken and shrimp are convenient sources of high quality low fat protein.
- Making food look fun can encourage your child to eat and enjoy them more. Try including colourful vegetables and cutting them into interesting shapes!
- Steamed egg can be served in main meals or as snack. Some kids may be fussy about meat due to its texture. Soft and silky smooth steamed eggs could be a more pleasant protein option.

Vegetable Okonomiyaki

| makes 3 servings | refer to p.75 |

Ingredients

50 g cake flour / 0.5 g bonito flakes / 25 g yam (grated) / 100 g shredded cabbage / 1 whole egg (medium) / salt / 1 tsp vegetable oil

Seafood fillings (optional)

scallop, shrimp or squid

Toppings

1 tbsp low fat mayonnaise / 1 tbsp okonomiyaki sauce / plain seaweed flakes / bonito flakes

Method

1. In a large mixing bowl, combine flour, bonito flakes and salt, mix well. (Kid-friendly step)
2. Stir in the grated yam, egg, shredded cabbage and seafood until combined. (Kid-friendly step)
3. In a small non-stick pan over medium heat, heat oil and pour in 1/2 portion of the batter. Cook the pancake for 5 minutes (flip midway), until fully cooked and lightly browned. Serve with both sauces (1 tbsp each), sprinkle on seaweed flakes and bonito flakes as desired. Serve.

Kid-friendly steps

Nutrition tips

- Substituting high fat pork belly slices from traditional okonomiyaki recipes with low fat seafood options would not only lower the overall fat intake but also add delicious seafood flavours.
- Cooking pancakes on non-stick pans could help control / limit the use of oil in recipes.

Rainbow Veggie Noodle Salad

| makes 2 servings | refer to p.78 |

Ingredients

80 g plain noodles / 1/4 red bell pepper / 1/4 yellow bell pepper / 1/4 green bell pepper / 2 okras / 4 cherry tomatoes / 4 baby corns / 4 frozen prawns / 4 frozen scallops / 4 tsp low fat sesame sauce / 1 tsp black sesame seeds / plain seaweed / Katsuobushi shavings

Method

1. Cook the noodles according to the packet instructions, set aside to cool.
2. Dice the bell peppers. Cut the okras into star shapes and cut the baby corns in half, blanch them and set aside. Cut the cherry tomatoes in half and set aside.
3. Defrost the frozen prawns and scallops. Blanch and cut into suitable shapes, set aside.
4. Plate the noodles with all cooked ingredients, serve with low fat sesame sauce, black sesame seeds, some plain seaweed and Katsuobushi shavings.

Kid-friendly steps

Nutrition tips

- Summer heat may reduce a child's appetite. This refreshing dish is perfect for a hot summer day.
- Frozen prawns and scallops are easily found and stored. Both are low in fat and rich in protein.
- Eating a variety of brightly coloured vegetables can help you get the full range of health benefits.
- When choosing low fat sesame sauce, remember to read the nutrition label, or you may replace it with vinegar or fresh lemon juice to further lower the fat and sodium intake.

Reindeer Beef Rolls

| makes 4 servings | refer to p.81 |

Ingredients

2 slices tortilla wrap / 8 pieces pretzel / 80 g beef fillet steak / 1/4 head lettuce (small) / 1/2 carrot / 1/2 red onion / 4 cherry tomatoes / 1/4 tsp grated garlic / 20 g mozzarella cheese / 1/2 avocado / 1/2 tsp fresh lemon juice / 1 tsp canola oil

Seasoning

1/8 tsp salt / 1/2 tsp sugar / 1/2 tsp cornflour / Italian herbs

Method

1. Rinse the beef fillet and pat dry with a kitchen towel. Season and let it marinate for 15 minutes.
2. Rinse the lettuce, carrot, red onion and cherry tomatoes. Tear the lettuce into pieces, finely slice the carrot and red onion, cut the cherry tomatoes in half and set aside.
3. In a small non-stick pan over medium heat, heat canola oil and add in red onions and the marinated beef until cooked through, whilst also adding minced garlic and stir-fry until fragrant.
4. Combine avocado and fresh lemon juice and mash. Break the pretzels in half.
5. Using the tortilla wrap as a base, spread a layer of avocado, then add in lettuce, carrot, red onion and beef accordingly. Lastly, add the mozzarella cheese.
6. Wrap up your creation and cut the roll into 4 equal pieces. Top with cherry tomatoes and pretzel pieces to create your reindeer roll!

Kid-friendly steps

Nutrition tips

- Beef is a source of protein and iron. Inadequate iron intake leads to iron deficiency anemia. Symptoms of iron deficiency anemia include pale skin, weakness, fatigue and poor concentration. It can also cause dizziness, loss of appetite, weakened immune system and other health problems.
- The recommended iron intake varies by age and gender, consult your doctor or registered dietitian. For example, a 4-6 years old child should consume 10mg of iron daily, regardless of their gender.
- Eggs, dried fruit, and dark green vegetables are also sources of iron. Plant-based iron requires vitamin C to boost iron absorption.

Greek Yogurt Pizza Crust

| makes 4 servings | refer to p.84 |

Ingredients - pizza base

110 g self-rising flour / 1/2 cup low fat plain Greek yogurt

Toppings

2-4 tbsp low sodium tomato sauce / 1/4 cup baby spinach / 1/4 cup sliced button mushroom / 1/4 cup cooked shredded chicken / 2 tbsp corn kernels / 1/4 cup cherry tomatoes (rinsed and halved) / 2 slices low fat cheese

Method

1. Preheat the oven to 200°C.
2. In a large mixing bowl, combine the ingredients of the pizza base, mix well and knead the dough until smooth (add extra flour as needed).
3. Transfer the dough to a lightly floured surface, divide the dough evenly into 2 separate dough balls, roll and flatten the dough into 9-inch rounds. Transfer the pizza base to a lined and floured baking tray.
4. Spread the tomato sauce evenly over the dough, sprinkle on the toppings and cheese at last. Let children create their very own pizza!
5. Bake it for around 20 minutes until the cheese has melted and the crust has turned slightly golden brown. Cut into slices and serve.

Kid-friendly steps

Nutrition tips

- Be creative with the pizza toppings, you may replace chicken with seafood such as shrimps or squids or even cooked fish! You could name the pizza after your child, praise their efforts.
- If you don't wish to make your own pizza dough, you may use the following as a pizza base: whole wheat pita bread, whole grain English muffins, whole wheat tortilla or even a slice of whole wheat toast.

Spinach and Tomato Lasagna

| makes 2 servings | refer to p.86 |

Ingredients

35 g lasagna sheets (no precooking needed) / 80 g baby spinach / 1/4 cup low sodium tomato sauce or homemade tomato sauce / 3 tbsp low fat cottage cheese / 2 tbsp parmesan cheese (grated) / 1/2 egg (medium-sized) / salt / ground black pepper

Method

1. Preheat the oven to 180°C. Cook spinach for a minute, squeeze dry and chop them into small pieces.
2. In a mixing bowl, combine the cooked baby spinach, low fat cottage cheese, egg, salt, pepper and mix well.
3. Spread about 2 tbsp of tomato sauce into a baking tray. Add a layer of lasagna sheet. Top the sheet with 1/2 portion of the spinach mixture. Repeat the layers. Top the last layer of lasagna sheet with the remaining tomato sauce and parmesan cheese.
4. Cover with the aluminium foil and bake it for around 30-40 minutes. Remove the foil and serve.

Kid-friendly steps

Nutrition tips

- Frozen vegetables are good alternatives to fresh vegetables, they are equally nutritious and convenient for some people. In this recipe, you may replace fresh spinach with frozen chopped spinach. Yet, when selecting frozen foods, don't forget to read the food label / ingredient list and avoid those with added salt, sugar and oil.
- Low fat cottage cheese can be used to make nutritious snacks, it is a source of protein and it can be used as a substitute for high fat creamy cheeses in some recipes.

Italian Gnocchi

| makes 8 servings | refer to p.88 |

Ingredients - gnocchi

450 g potato (steamed and mashed) / 1 cup all purpose flour / 1/4 tsp salt

Tomato sauce

1 onion (medium-sized, diced) / 1 tsp grated garlic / 1 tbsp olive oil / 1 can low sodium crushed tomato / 1/2 - 1 tsp Italian herbs (no added salt) / salt

Condiments

1/4 onion (medium-sized, diced) / 1/4 cup cherry tomatoes (rinsed, cut into half) / 3 tbsp parmesan cheese / 1 tbsp olive oil / fresh basil

Method

1. Prepare the tomato sauce by heating up the olive oil on a non-stick pan, cook the diced onion over low heat until soft. Add in the remaining ingredients and let it simmer for around 20 minutes. Stir occasionally, once it's done, let it cool and set aside.
2. In a large mixing bowl, add in the mashed potato, flour and salt. Mix well until a dough is formed.
3. On a lightly floured work surface, keep kneading the dough and divide them into 8 equal portions. Roll them into ropes and cut them into small pieces (the size of your thumb), roll each piece on the back of a metal fork to create that signature ridges.
4. Prepare some boiling water, boil the gnocchi until they float, drain them thoroughly.
5. In a heated non-stick pan, add in some olive oil, slowly pan fry the onion and gnocchi for around 3 minutes or until golden, add in the cherry tomatoes and 1-2 cups of tomato sauce. Once it's ready, add in the fresh basil and parmesan cheese, serve.

Kid-friendly steps

Nutrition tips

- Tomatoes not only give us great flavours but are a great source of vitamin C, A and lycopene. Heating tomatoes with some oil may help boost the absorption of lycopene.
- Both dried and fresh herbs are flavourful and are wonderful to use in cooking. They add both flavours and nutrients to our dishes, allowing us to use less salt in cooking! Basil, for example, works especially well with tomato and pasta dishes.
- To make the gnocchi more colourful, feel free to add in some spinach and beet puree to the gnocchi recipe.

Rose Dumplings

| makes 8 pieces/servings | refer to p.90 |

Ingredients

300 g lean minced pork / 1 tbsp instant oat / 2 tbsp frozen mixed vegetables / 24 dumpling wrappers / water

Seasoning

1 tsp sesame oil / 1/4 tsp salt / 1/4 tsp sugar / 1/2 tsp soy sauce

Method

1. Marinate the minced pork with the seasoning for 15 minutes.
2. Defrost the frozen mixed vegetables and chop them into small pieces.
3. In a large bowl, mix the marinated pork, frozen mixed vegetables and oats together. Wrap in cling film and refrigerate for 15 minutes.
4. Take out three dumpling wrappers and lay them out horizontally, overlapping 1/3 of each wrapper and stick them together with water.
5. Scoop around 1 tbsp of dumpling filling and place it onto the centre of the dumpling wrapper. Fold the dumpling wrapper from bottom to top into a semi circle shape. Rub some water on the edge to stick it. Roll up the dumpling wrappers from left to right into a rose shape, using water to stick the ends together.
6. In a large pot, bring water to a boil and steam dumplings for 15 minutes. Serve.

Kid-friendly steps

Nutrition tips

- Instant oats can absorb excess water from the dumpling filling, while providing soluble dietary fibre.
- Remember to buy lean pork to keep your saturated fat intake low.
- Uncooked dumplings can be stored in the freezer for several days, you can simply steam and serve them with vegetables or noodles as a quick and easy meal.

Seaweed Chicken Rolls

| makes 5 servings | refer to p.92 |

Ingredients

2 sheets of sushi nori (each sheet sliced into 10 even rectangular pieces) / 1 tsp vegetable oil

Chicken fillings

200 g chicken breast (minced) / 180 g cabbage (finely chopped and squeeze dried) / 20 g onion (finely diced) / 1/2 tsp salt / ground white pepper / 1 tsp sesame oil

Method

1. In a large bowl, combine all the ingredients of chicken fillings. Mix well with chopsticks, until well combined, and a slightly sticky patty is formed.
2. Evenly spread around 20 g of chicken filling on each small piece of sushi nori, and roll up.
3. In a heated pan, add around 1 tsp of vegetable oil, pan fry them over low heat until they are slightly golden brown and cooked through. Serve.

Kid-friendly steps

Nutrition tips

- Chicken breast is a relatively lean cut, we could add more texture, flavour and nutrients to the meat by grinding them and mixing in some finely chopped vegetables.
- Be creative and try mixing in all sorts of vegetables such as carrots, radish, spring onion, mushrooms and even corn kernels. Allow children to try a variety of vegetables to ensure that they are getting a wide range of nutrients!

Princess Omurice Gown

| makes 4 servings | refer to p.94 |

Ingredients

100 g minced beef / 1/4 onion / 1 tsp minced garlic / 5 button mushrooms / 200 g canned tomato puree / 50 g frozen mixed vegetables / 1 bowl rice or oat rice / 2 whole eggs / 1 tbsp low fat milk / 3 tsp canola oil / 1/4 tsp sugar / 1/4 tsp salt

Seasoning

1 tsp soy sauce / 1 tsp cornflour / 1/2 tsp sugar

Method

1. Marinate the minced beef with the seasoning for 15 minutes, set aside.
2. Dice the onion, slice the button mushroom, rinse and pat dry the mixed vegetables, set aside.
3. Heat up the wok and add 2 tsp of oil. Stir-fry the onion until it softens, add garlic and stir-fry until fragrant. Add the button mushroom and minced beef, stir-fry until 80% cooked, add tomato puree and cook over low heat until the sauce thickens.
4. Add the mixed vegetable, rice, sugar and salt and stir-fry until well combined. Pack the rice into two bowls and place a plate on top and flip both over.
5. Shape the rice into a gown shape with cling film.
6. Whisk together the egg and milk, pass the mixture through a sieve.
7. Heat a pan over medium heat, add 1 tsp of oil and pour in the egg mixture. Swirl around and evenly coat the bottom of the pan. Cover the pan and turn off the heat, let the egg cook completely through.
8. Place the egg onto the fried rice; decorate the dish as shown with vegetables and props.

Kid-friendly steps

Nutrition tips

- Eggs and beef both provide quality protein. One egg equates to one meat portion. Eggs are affordable, readily available and are easy to chew.
- When choosing canned tomato puree, pay attention to the ingredient list and nutrition label, choose products that are low in sugar and sodium.
- Minced beef in this recipe can be replaced by skinless chicken fillet, red kidney beans or other protein sources. Get creative and decorate your dish with vegetables of your choice!

Tri-colour Fruity Sorbet

| makes 4 servings | refer to p.97 |

Ingredients - melon sorbet

1 cup melon chunks / 1 tsp lime juice / 1 tsp maple syrup

Ingredients - honeydew melon sorbet

1 cup honeydew melon chunks / 1 tsp lime juice / 1 tsp maple syrup

Ingredients - watermelon sorbet

1 cup watermelon chunks / 1/2 ripe banana / 1 tsp lime juice

Method

1. Put melon chunks, honeydew melon chunks and watermelon chunks into separate ziplock bags and freeze for 4-6 hours. 🧑‍🍳
2. Combine melon chunks, lime juice and maple syrup in a blender and blend until smooth. 🧑‍🍳
3. Combine honeydew chunks, lime juice and maple syrup in a blender and blend until smooth. 🧑‍🍳
4. Combine watermelon chunks, banana and lime juice in a blender and blend until smooth. 🧑‍🍳
5. Pour the mixture back into the 3 ziplock bags, freeze for 1 hour, take it out and gently massage the bags for about 2 minutes. Freeze them for 1 more hour, serve and enjoy. 🧑‍🍳

🧑‍🍳 Kid-friendly steps

Nutrition tips

- Melons are not only delicious and refreshing but also contain natural fructose, rich in water, low in fat and are packed with nutrients.
- A cup of melon (~177 g) provides around 64 kcal and 1.4 g of dietary fibre. It is a source of nutrients such as vitamin C, B_6 and potassium. A cup of honeydew melon (~177 g) provides around 60 kcal and 1.6 g of dietary fibre. It contains a range of antioxidants such as beta carotene, lutein and zeaxanthin. Watermelon is also great, each cup of watermelon (~154 g) provides 47 kcal and 0.6 g of dietary fibre, it is a good source of vitamin C, beta carotene and lycopene.

Baked Apple Turnover

| makes 4 servings | refer to p.100 |

Ingredients

4 crustless sandwich bread (large, preferably whole wheat) / 1 whole egg / 1 tbsp low fat milk

Fillings

80 g apples (peeled and diced)
1 tsp brown suga
2 tbsp-1/4 cup drinking water
1 tsp cornflour / 1/4 tsp ground cinnamon
1/2 tsp vanilla extract

Method

1. Add eggs and low fat milk to a large bowl, mix well and set aside.
2. Add all ingredients for the fillings (excluding apples) into a small pot, mix well and bring it to a boil. Add diced apple and cook over low heat until the apples soften.
3. Flatten the bread with a rolling pin, cut 3 slashes on one side of the bread. Spread an appropriate amount of egg wash on the edge, and add in about 20 g of apple fillings.
4. Fold the bread in half, seal the side with a fork and lightly coat them with egg wash.
5. Bake them in a preheated oven at 180°C for 10 minutes or until slightly golden brown, or air-fry it for about 4 minutes. Serve.

Kid-friendly steps

Nutrition tips

- Puff pastry is often used as the main ingredient in most tarts and pies. In most cases, butter or shortening is used when making puff pastry, resulting in a higher content of unhealthy fats such as saturated fat or even trans fat. Replacing puff pastry with whole wheat bread would not only make the apple pie healthier but also easier for kids to handle.
- Other healthy and delicious fillings combos may include blueberry and low fat yogurt or banana and peanut butter etc.

Dreamy Smoothie

| makes 2 servings | refer to p.102 |

Ingredients - first layer

1/4 cup low fat yogurt / 1/4 cup low fat milk / 1/2 cup blueberries (frozen) / 1/2 ripe banana / 1/2 tbsp chia seeds

Ingredients - second layer

1/4 cup low fat yogurt / 1/4 cup low fat milk / 1/2 cup mango chunks (frozen) / 1/2 ripe banana / 1/2 tbsp chia seeds

Method

1. In a blender, combine all ingredients for the first layer and blend until smooth, divide the smoothie between 2 cups (as the first layer).
2. Wash the blender and then combine all ingredients for the second layer in the blender and blend until smooth, divide them evenly into the cups as the second layer. Stir slowly with a straw and enjoy.

Kid-friendly steps

Nutrition tips

- Chia seeds are rich in Omega-3 fatty acids, dietary fibre, protein, and various minerals such as calcium, iron, magnesium, and zinc. Also, 15 g of chia seeds provides around 5 g of dietary fibre.

Mixed Fruit Jelly Brick

| makes 4 servings | **refer to p.104 |**

Ingredients

150 g melon / 150 g honeydew melon / 15 g sugar / 200 ml water / dried butterfly pea flower / 16 g gelatin powder / 100 ml low fat milk / mint leaves

Method

1. Scoop small spheres out of the melon and honeydew melon. Mix and transfer to a container. 🧑‍🍳
2. Add sugar and a small amount of butterfly pea flower into the water and heat until the sugar dissolves, the mixture should turn blue and by adding a small amount of lemon juice you can turn the mixture purple. 🧑‍🍳
3. Turn off the heat and add 10 g of gelatin powder; stir until it is completely dissolved. Pour into the container (with fruits) and refrigerate.
4. Heat up the low fat milk over low heat, add the remaining 6 g of gelatin powder and mix well. Pour the mixture onto the solidified fruit jelly.
5. Refrigerate until the milk layer is set. Cut them into 1-inch bricks and serve. 🧑‍🍳

🧑‍🍳 Kid-friendly steps

Nutrition tips

- Dried butterfly pea flower provides distinctive and attractive colours to the dish, which can enhance children's interest in food and the science behind it.
- This recipe can be adapted to suit your child's fruit preference. This is a healthier substitute for store bought jellies.
- Any fresh fruit can be a healthy choice, however, avoid using fresh pineapple for homemade jelly. Pineapple contains the enzyme bromelain which can break down the protein in gelatin powder, preventing gelatin to set.
- Canned pineapple can be used instead, yet choose fruit that is canned in water or in its own juice.

Fruity Yogurt Parfait with Chia Seed Jam

| makes 2 servings | **refer to p.107 |**

Ingredients - chia seed jam

80 g strawberries / 2 g honey / 5 g chia seeds / 3 g fresh lemon juice / fresh lemon peel

Method

1. Rinse and dice the strawberries, heat them in a small saucepan over low heat, stir occasionally, until softened.
2. Mash the berries to your desired consistency.
3. Turn off the heat and stir in honey, fresh lemon juice, lemon peel and chia seed.
4. Extra chia seed may be added as desired, the jam will thicken as it cools.

Ingredients - fruity yogurt parfait

1 small cup low fat plain yoghurt / 1 kiwi / 2 strawberries / 1/2 box blueberry / 1 mandarin /
1/2 box blackberry / 4 tbsp corn flakes (low fat and low sugar) /
4 tsp homemade strawberry chia seed jam / mint leaves (for decoration)

Method

1. Dice or slice the fruits and set aside.
2. In a serving glass, layer cornflakes, yoghurt, fruits and jam as desired. Repeat the process until the glass is filled. Top it up with mint leaves and serve.

Kid-friendly steps

Nutrition tips

- Fresh fruit and chia seeds provide dietary fibre, which is essential for our health, it could help to keep our digestive system healthy and prevent constipation.
- Homemade chia seed jam is a healthy and delicious alternative to sugar-packed store-bought jam.
- Low fat plain yogurt provides protein and calcium and contains potassium.

Mango Soy Milk Pudding

| makes 2 servings | refer to p.110 |

Ingredients

250 ml low sugar calcium fortified soy milk / 3.5 g gelatine powder / 1/4-1/2 cup diced mango

Method

1. Place the soy milk in a medium-sized saucepan. Sprinkle gelatin over and let it sit for 5 minutes. Place the saucepan over low heat, keep stirring until gelatin dissolves. Do not allow the mixture to boil. Turn off the heat.
2. Pass the mixture through a sieve, pour the mixture evenly in 2 small jars, cover the jars and let It refrigerate for around 6 hours or overnight.
3. Mash the mango with a fork, top the soy milk pudding with your desired amount of mango puree. Serve.

Kid-friendly steps

Nutrition tips:

- Soy milk is a source of protein, but is naturally low in calcium. If you wish to replace cow's milk with soymilk, choose calcium fortified soy milk.
- Mango is naturally sweet and juicy, it is great for dessert and no added sugar is needed. Usually, the riper the mango, the sweeter it is.

Banana Oat Animal Pancake

| makes 2 servings | refer to p.112 |

Ingredients

1 ripe banana / 1 whole egg / 1/4 cup oats / 1/8 tsp baking powder / 1/4 tsp ground cinnamon / 1/8 tsp salt / 1 tsp canola oil / 1 cup low fat plain yogurt / fresh fruits / plain nuts

Method

1. Place the banana, egg, oats, baking powder, cinnamon and salt into a blender and blend until smooth.
2. Heat up a non-stick pan over medium heat, add oil, then wipe off excess with a kitchen towel. Add 1/2 or 1/4 pancake batter and fry until both sides are golden brown.
3. Let it cool, decorate them with low fat plain yogurt, fresh fruit and plain nuts. Serve. (Children can use cookie cutters or blunt plastic knives to cut out different shapes of fruits for decoration.)

Kid-friendly steps

Nutrition tips

- These blender pancakes are easy to make and are good substitute for store-bought pancake mixes that are higher in added sugar.
- Oats offer soluble fibre, protein and a range of vitamins and minerals. Serving these pancakes with colourful fruits would make the dish even more attractive and nutritious.

Foolproof Christmas Appetizers

| makes 2 servings | refer to p.114 |

Ingredients

1 slice (small) whole wheat pita bread / 4 tbsp guacamole (refer to p.127 recipe)

For decoration

tri-colour bell pepper (diced) / high fibre cereal sticks

Method

1. Divide the pita bread into 8 equal wedges. Spread around 1-1.5 tsp of guacamole on each wedge.
2. Remove any excess water from the diced bell peppers with a paper towel.
3. Decorate the Christmas tree with the diced bell peppers and high fibre cereal sticks (bell peppers as ornaments and cereal as the tree trunk).

Kid-friendly steps

Nutrition tips

- Whole wheat pita bread is light and great for many recipes. Not only does it go well with guacamole or hummus, but it could also be used as a simple pizza base for DIY pizzas!
- You may also prepare sandwiches with whole wheat pita bread, simply cut it in half and fill the pocket with desired sandwich fillings. It's a fun and simple lunch option.
- Cherry tomatoes or other fresh fruits and vegetables can be used to decorate the Christmas tree.

The Perfect Energy Balls for Chinese New Year

| makes 10 pieces/servings | refer to p.116 |

Ingredients

4 tbsp rolled oats / 2 tbsp all-natural peanut butter / 1 tbsp pistachio / 1 tbsp almonds / 1 tbsp dried cranberries or unsweetened raisins / 1 tbsp goji berries / 1 tsp honey / 1 tsp ground cinnamon (optional)

Method

1. Roughly chop the pistachios, almonds, dried cranberries and goji berries. Save some whole for decoration.
2. In a large bowl, combine rolled oats, pistachios, almonds and cinnamon. Add in the peanut butter, honey, dried cranberries, goji berries and mix well.
3. Place 1 tbsp of the mixture onto a piece of cling film, roll them into balls and serve.

Kid-friendly steps

Nutrition tips

- Soluble fibre found in oats can help lower blood cholesterol and glucose levels.
- All-natural peanut butter with no added salt, sugar and oil is a healthier option.
- Nuts are healthy snack options, they're good sources of healthy fat, fibre and protein.

Pumpkin Chinese New Year Treats

| makes 12 pieces/servings | refer to p.118 |

Ingredients

150 g pumpkin (diced) / 15 g cane sugar / 1/8 tsp salt / 75 g glutinous rice flour / pumpkin seeds / high fibre cereal sticks (for decoration)

Fillings

25 g bananas (ripe) / 25 g crushed unsalted peanuts / 15 g cane sugar

Method

1. Steam the pumpkin pieces until tender, remove excess water if needed.
2. Use a fork to mash the pumpkin until smooth, mix in cane sugar, salt and glutinous rice flour, knead the mixture until a dough forms.
3. Prepare the fillings by mixing in the mashed banana, crushed peanuts and cane sugar.
4. Divide the dough into 12 equal portions, shape them into small balls. Flatten a piece of dough into a round wrapper with your hands, place 1 tsp of the fillings in the middle, gently seal the wrapper. Shape the dough and pierce the surface as shown with toothpicks.
5. Steam the treats over medium heat for 10-12 minutes. Assemble the treat with pumpkin seeds and high fibre cereal sticks, serve.

Kid-friendly steps

Nutrition tips

- Pumpkin seeds are a great source of unsaturated fat and are rich in nutrients such as iron, zinc and magnesium.

Shiba Inu Glutinous Rice Balls

| makes 5 servings | refer to p.120 |

Ingredients

90 g glutinous rice flour / 100 g tofu / 2 tbsp soy sauce / 15 g dark brown sugar / 50 ml water / 1 tsp cornflour (mixed with water) / black sesame seeds

Method

1. Combine glutinous rice flour and tofu. Save up a small portion of the dough (for the ears, eyebrows, nose and tail), divide the remaining into 10 equal portions, roll them into balls.
2. Form your shiba inu ears, eyebrows, nose and tail with the saved up dough. Stick them onto the balls using tiny amounts of water, to create a shiba inu head or body.
3. Gently boil them until the balls float, set aside.
4. Add soy sauce, dark brown sugar and water in a small pot, dissolve the sugar over low heat. Add cornflour slurry and cook until the sauce thickens.
5. Brush the sauce onto the shiba inu and decorate them with black sesame seeds, serve.

Kid-friendly steps

Nutrition tips

- Tofu is a source of protein and is used to replace some glutinous rice flour in this recipe.
- Most of the glutinous rice ball found in the market may contain palm oil, a source of unhealthy saturated fat.

A-Z 食物營養字咕，邊吃邊學營養食材！

與父母動動手下廚之後，不妨在色彩繽紛的食物中，透過可愛的 A-Z 食物營養字咕認識不同的食物，來一趟食物營養之旅吧！

檸檬
Lemon

牛奶
Milk

果仁
Nuts

橙
Orange

菠蘿
Pineapple

藜麥
Quinoa

飯糰
Rice Ball

士多啤梨
Strawberry

番茄
Tomato

烏冬
Udon

雲呢拿
Vanilla

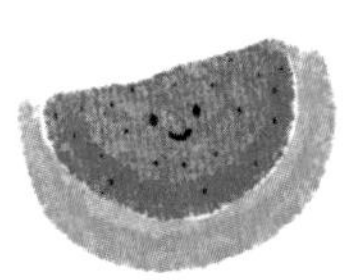
西瓜
Watermelon

小籠包
Xiao Long Bao

乳酪
Yogurt

翠玉瓜
Zucchini

著者
陸蕙華（Denise Luk）、盧庭威（Kurtus Lo）

責任編輯
簡詠怡

裝幀設計
鍾啟善

攝影
Tango Chan、Steven Wu

排版
辛紅梅、何秋雲

出版者
萬里機構出版有限公司
香港北角英皇道 499 號北角工業大廈 20 樓
電話：2564 7511　　傳真：2565 5539
電郵：info@wanlibk.com
網址：http://www.wanlibk.com
http://www.facebook.com/wanlibk

發行者
香港聯合書刊物流有限公司
香港荃灣德士古道 220-248 號荃灣工業中心 16 樓
電話：2150 2100　　傳真：2407 3062
電郵：info@suplogistics.com.hk
網址：http://suplogistics.com.hk

承印者
美雅印刷製本有限公司
香港九龍觀塘榮業街 6 號海濱工業大廈 4 樓 A 室

出版日期
二〇二一年四月第一次印刷

規格
特 16 開（240mm × 170mm）

ISBN 978-962-14-7325-7